T0181618

How to Write Technical Reports

Heike Hering

How to Write Technical Reports

Understandable Structure, Good Design, Convincing Presentation

Second Edition

 Springer

Heike Hering
Hannover, Niedersachsen, Germany

ISBN 978-3-662-58105-6   ISBN 978-3-662-58107-0   (eBook)
https://doi.org/10.1007/978-3-662-58107-0

Library of Congress Control Number: 2018957060

Original German edition published by Vieweg, 2007
1st edition: © Springer-Verlag Berlin Heidelberg 2010
2nd edition: © Springer-Verlag GmbH Germany, part of Springer Nature 2019

This Springer imprint is published by the registered company Springer-Verlag GmbH, DE part of Springer Nature
The registered company address is: Heidelberger Platz 3, 14197 Berlin, Germany

Preface

Technical Reports are usually written according to general standards, corporate design standards of the current university or company, logical rules, and practical experiences. These rules are not known well enough among writing engineers and technicians. Therefore, this book intends to help you create Technical Reports. It contains many practical examples. It is based on a German edition published by Springer Vieweg and is now published by Springer as second edition in English.

Both authors of the 1st to 7th German edition have long experience in educating engineers at the University of Applied Sciences Hannover. They have held many lectures where students had to write reports and took notes about all positive and negative examples that occurred in design reports, laboratory work reports, and in theses. Dr. Heike Hering worked at TÜV NORD Akademie, where she was responsible for e-learning projects and Internet pages and supervised students who were writing their theses. She now works for her private teaching company.

Prof. Dr.-Ing. Klaus-Geert Heyne joined the team as co-author for the second German edition. He redesigned Chap. 5 "Presenting the Technical Report." He contributes his experiences from Motorenwerke Mannheim AG and his lectures at University of Applied Sciences Wiesbaden Rüsselsheim as well as from rhetoric and visualization seminars in Rüsselsheim and Mannheim.

This book answers questions of engineering students and practitioners occurring when writing Technical Reports or preparing presentations on the PC. These questions refer to contents as well as formal aspects. Such questions occur during the whole work on the report or presentation from the beginning to the end. Therefore, this book is designed as a guideline or manual *How to Write Technical Reports*. It is ordered by timeline along the process of writing Technical Reports into the three phases: planning, creation, and finishing.

In this second edition, the book was changed to fit to an electronic and media neutral production process, so that an e-book version can easily be created that runs on computers and mobile devices. The following sections have been shortened a lot: Tables (Sect. 3.3), figures (Sect. 3.4), word processing and desktop publishing systems (Sect. 3.7), and presentation graphics programs (Sect. 5.4.3). The section about copyright and copyright laws has been deleted. The topic short statement (Sect. 5.7) has been added.

Hannover, Germany   Heike Hering
September 2018

Contents

1 Introduction   1

2 Planning the Technical Report   5
   2.1 General Overview of All Required Work Steps and Time Planning   5
   2.2 Accepting and Analyzing the Task   7
   2.3 Checking or Creating the Title   8
   2.4 The Structure as the "Backbone" of the Technical Report   11
      2.4.1 General Information About Structure and Table of Contents   12
      2.4.2 Rules for the Structure in ISO 2145   12
      2.4.3 Logic and Formal Design of Document Part Headings   14
      2.4.4 Work Steps to Create a Structure and Example Structures   18
      2.4.5 General Structure Patterns for Technical Reports   25
   2.5 Project Notebook (Jotter)   29
   2.6 The Style Guide Advances Consistency in Wording and Design   30
   2.7 References in This Chapter   32

3 Writing and Creating the Technical Report   33
   3.1 Parts of the Technical Report and Their Layout   34
      3.1.1 Front Cover Sheet and Title Leaf   35
      3.1.2 Structure with Page Numbers = Table of Contents (ToC)   41
      3.1.3 Text with Figures, Tables, and Literature Citations   47
      3.1.4 List of References   49
      3.1.5 Other Required or Useful Parts   49
   3.2 Collecting and Ordering the Material   56
   3.3 Creating *Good* Tables   58
      3.3.1 Table Numbering and Table Headings   58
      3.3.2 The Morphological Box—A *Special* Table   61
      3.3.3 Hints for Evaluation Tables   65
      3.3.4 Tabular Re-arrangement of Text   69
   3.4 Instructional Figures   70
      3.4.1 Understandable Design of Instructional Figures   72

3.4.2   Figure Numbering and Figure Subheadings 75

3.4.3   Scheme and Diagram (Chart)........................... 77

3.4.4   The Sketch as Simplified Technical Drawing
and Illustration of Computations 82

3.4.5   Perspective Drawing 84

3.4.6   Technical Drawing and Bill of Materials (Parts List)......... 85

3.4.7   Mind Map 87

3.4.8   Pictorial Re-arrangement of Text 88

3.4.9   Creating Paper Images and Graphics Files and Incorporating
Them into the Technical Report....................... 88

3.5   Literature Citations 96

    3.5.1   Introductory Remarks on Literature Citations 96

    3.5.2   Reasons for Literature Citations 97

    3.5.3   Bibliographical Data According to ISO 690 98

    3.5.4   Citations in the Text 98

    3.5.5   The List of References—Contents and Layout 107

3.6   The Text of the Technical Report 119

    3.6.1   Good Writing Style in General Texts 119

    3.6.2   Good Writing Style in Technical Reports 120

    3.6.3   Formulas and Computations 123

    3.6.4   Understandable Writing in Technical Reports 127

3.7   Using Word Processing and Desktop Publishing (DTP) Systems 132

3.8   Completion of the Technical Report 140

    3.8.1   The Report Checklist Assures Quality and Completeness 140

    3.8.2   Proof-Reading and Text Correction According to ISO 5776 142

    3.8.3   Creating and Printing the Copy Originals and End Check 146

    3.8.4   Exporting the Technical Report to HTML or PDF
for Publication 149

    3.8.5   Copying, Binding or Stapling the Technical Report
and Distribution 151

3.9   References in This Chapter................................... 158

4   **Useful Behavior for Working on Your Project and Writing
the Technical Report** 159

4.1   Working Together with the Supervisor or Customer 160

4.2   Working Together in a Team 161

4.3   Advice for Working in the Library 162

4.4   Organizing Your Paperwork................................. 164

4.5   Organizing Your File Structure and Back-up Copies 165

4.6   Personal Working Methodology 168

4.7   References in This Chapter................................... 171

5   Presenting the Technical Report 173
  5.1   Introduction 173
      5.1.1   Target Areas University and Industrial Practice 173
      5.1.2   What Is It All About? 174
      5.1.3   What Is My Benefit? 175
      5.1.4   How Do I Proceed? 176
  5.2   Why Presentations? 176
      5.2.1   Definitions 176
      5.2.2   Presentation Types and Presentation Targets 177
      5.2.3   "Risks and Side Effects" of Presentations and Lectures 178
  5.3   Planning the Presentation 180
      5.3.1   Required Work Steps and Their Time Consumption 180
      5.3.2   Step 1: Defining the Presentation Framework and Target 181
      5.3.3   Step 2: Material Collection 186
      5.3.4   Step 3: The Creative Phase 187
  5.4   Creating the Presentation 191
      5.4.1   General Recommendations for Designing Presentation Slides ... 192
      5.4.2   Step 4: Summarizing the Text and Working Out the Details 196
      5.4.3   Step 5: Visualization and Manuscript 197
      5.4.4   Step 6: Trial Presentation and Changes 210
      5.4.5   Step 7: Updating the Presentation and Preparations
            in the Room 211
      5.4.6   Step 8: Lecture, Presentation 212
  5.5   Giving the Presentation 212
      5.5.1   Contact Preparations and Contacting the Audience 213
      5.5.2   Creating a Relationship with the Audience 214
      5.5.3   Appropriate Pointing 215
      5.5.4   Dealing with Intermediate Questions 215
  5.6   Review and Analysis of the Presentation 216
  5.7   The Short Statement 219
      5.7.1   Trigger 219
      5.7.2   Requirements 220
      5.7.3   Example Statement for a Familiar Audience 220
      5.7.4   Example Statement for an Unfamiliar Audience 221
      5.7.5   Problems of Short Statements 221
      5.7.6   Tactical Measures 222
  5.8   57 Rhetoric Tips from A to Z 222
  5.9   References in This Chapter 226

6   Summary 227

Glossary—Terms of Printing Technology 229

Cited and Recommended References 235

Index 241

About the Author

Dr. Heike Hering has a private teaching institute in Hannover and has been a Lecturer at the Hochschule Hannover—University of Applied Sciences and Arts as well as FH Aachen for many years. She has written and translated software manuals for a CAD company in German and English and was a safety engineer for 9 years. Thereafter, she has worked at the TÜV NORD Academy in Hannover for 18 years and has developed e-learning programs, took part in software projects and worked as safety engineer. Prof. Dr.-Ing. Klaus-Geert Heyne represented the fields of communication, combustion engines, and statics at the Hochschule RheinMain University of Applied Sciences in Rüsselsheim.

List of Figures

Fig. 2.1   Network plan for creating Technical Reports   6

Fig. 2.2   Network plan to write Technical Reports—analysis of the task   8

Fig. 2.3   Network plan to write Technical Reports—create the title   8

Fig. 2.4   Network plan to write Technical Reports—creating the 4-point
and 10-point structure.   11

Fig. 3.1   Network plan to write Technical Reports—search and cite literature,
create text, figures and tables, develop a detailed structure.   34

Fig. 3.2   Comparison of a faulty (left side) and a correct (right side)
front cover sheet for a design report.   36

Fig. 3.3   Two handwritten drafts of the title leaf of a dissertation (the
placement of information varies between centered, left justified,
along a line, and right justified)   38

Fig. 3.4   Two handwritten drafts of the title leaf of a dissertation
(with variation of font size and line breaks)   38

Fig. 3.5   Front cover sheet and title leaf of a diploma thesis   39

Fig. 3.6   Front cover sheet and title leaf of a design report   39

Fig. 3.7   Example of a chapter ToC for an Appendix.   55

Fig. 3.8   Technical-economical evaluation of the concept variants
of a water purification plant (from VDI 2222)   67

Fig. 3.9   Layout of figure subheadings that spread across more than one line
and common figure subheading of several small figures.   76

Fig. 3.10   Diagram with physical values, measuring units and measures   78

Fig. 3.11   Example of a diagram with ruled lines to read off exact values.
Source Table appendix for the textbook ROLOFF/MATEK
Machine Elements   79

Fig. 3.12   Two variants to show that a scale or coordinate axis is interrupted . . .   79

Fig. 3.13   Different, clearly distinguishable measured point symbols
and line styles.   80

Fig. 3.14   Marking of the measured values and the limits of an error
tolerance zone within the true value lies (the two left variants
give the best contrast).   80

Fig. 3.15   Example for a curve diagram that shows only the qualitative
            relationship of two physical values (stress-strain-diagram
            of a tensile test)   80
Fig. 3.16   Change of the optical impression of a curve by change
            of the scale density.....................................   81
Fig. 3.17   Arm of a puller with diagram of forces and moments   83
Fig. 3.18   Gearbox scheme with exact specification of bearings, shafts
            and gears   83
Fig. 3.19   Examples of simplified technical drawings from manufacturer
            documents   84
Fig. 3.20   Simplified display of the sections of different semi-manufactured
            bar materials...   84
Fig. 3.21   Advantage of perspective drawings: the shape of objects can be
            determined much easier   85
Fig. 3.22   Avoidance of frequent mistakes in technical drawings   86
Fig. 3.23   Example of a mind map for planning the building of a pond
            in a garden   87
Fig. 3.24   Examples of pictorial re-arrangements of text....................   88
Fig. 3.25   Section through a protective gas welding pistol with integrated weld
            smoke suction (fill patterns and arrows show the direction of flow of
            protective gas and weld smoke)   91
Fig. 3.26   Components of a literature citation............................   99
Fig. 3.27   Layout of the list of references in three-column
            and two-column form.....................................   108
Fig. 3.28   Page layout with usual positions of the page number (positions a and
            b are recommended) and minimum size of the page margins........   133
Fig. 3.29   Network plan for the creation of Technical Reports: proofreading   143
Fig. 3.30   Simplified system of correction symbols according to ISO 5776   145
Fig. 3.31   Network plan for the creation of Technical Reports: printing originals
            or PDF and end check   146
Fig. 3.32   Network plan for creating Technical Reports: copy, bind
            and distribute the report   151
Fig. 3.33   Folding of a swing-out to the bottom.........................   152
Fig. 3.34   Folding a DIN A3 drawing or a swing-out to the right
            to DIN A4   153
Fig. 3.35   Folding a DIN A1 drawing to DIN A4   153
Fig. 5.1   Network plan to create and present a lecture   181
Fig. 5.2   Gantt diagram to create and present a lecture...................   181
Fig. 5.3   Processes compared with a presentation (biology)   189
Fig. 5.4   Processes compared with a presentation (technology)............   190
Fig. 5.5   Dramaturgy model of the knowledge level.....................   191
Fig. 5.6   Transparency assurance by a structure in a sidebar (schematic)   194
Fig. 5.7   Title slide (slide 1 of the example presentation)................   201
Fig. 5.8   Structure slide (slide 2 of the example presentation)   202

Fig. 5.9   Introduction slide (Slide 3 of the example presentation),
            "Well-known".. 202
Fig. 5.10   Causes of welding pollutants (slide 4 of the example presentation),
            "Well-known".. 203
Fig. 5.11   Effects of the welding pollutants (slide 5 of the example
            presentation), "New" 203
Fig. 5.12   Suction methods (slide 6 of the example presentation), "New" 204
Fig. 5.13   Original suction nozzle (slide 8 of the example presentation),
            "New" 205
Fig. 5.14   Optimized suction nozzle (slide 9 of the example presentation),
            "New" 205
Fig. 5.15   Measuring the isotaches (slide 11 of the example presentation),
            "Mazy".. 206
Fig. 5.16   Results of weighing the weld smoke (second last slide of the example
            presentation), "New" 207
Fig. 5.17   Summary (last slide of the example presentation),
            "Nothing New!"...................................... 207
Fig. 5.18   Manuscript card 209
Fig. 6.1   Network plan for creating a technical report: All working
            items are finished 227

List of Tables

Table 3.1   Morphological box (for a fast turn-off device in a nuclear power plant) with several concept variants, the electrical solution has been selected 63

Table 3.2   Morphological box (for a fast turn-off device in a nuclear power plant) with one concept variant 63

Table 3.3   Morphological box: sub function with several subgroups or characteristics. 64

Table 3.4   Morphological box: subdivided sub function 65

Table 3.5   Technical evaluation properties of a water purification plant (acc. to VDI 2222). 66

Table 3.6   Economical evaluation properties of a water purification plant (acc. to VDI 2222). 66

Table 3.7   Example of a technical evaluation table for the chassis of a boat trailer 68

Table 3.8   Systematic structure of graphic displays according to function and contents. 72

Table 3.9   Diagram types and their fields of application. 81

Table 3.10   Popular citation styles in mathematics, natural sciences, engineering and social sciences (each institution have their own variation of typography). 110

Table 3.11   Author specification in the list of references 110

Table 3.12   Usual abbreviations of terms in bibliographical data of publications 111

Table 3.13   Examples for the bibliographical data of common publication types in Springer format 112

Table 3.14   Data structure for bibliographical data of common publication types according to ISO 690 and example entries 112

Table 3.15   Dealing with information from data networks 116

Table 3.16   Example of a three-column list of references. 117

Table 3.17   Example of a space-saving list of references 118

Table 3.18   Examples for constituents of formulas 124

Table 5.1   Presentation types 178

Table 5.2   Advantages and disadvantages of written and oral
             communication.. 179
Table 5.3   Time consumption to create and present a lecture 182
Table 5.4   Third-rule (trisection) to design a presentation.................. 189
Table 5.5   Design of a technical presentation 190
Table 5.6   Trisection and assignment of contents in the example
             presentation 192
Table 5.7   Structure design plot for 20–60 min presentation time............ 196
Table 5.8   Evaluation scheme for a technical presentation with weighing
             the presentation elements................................ 216

Introduction

1

A scientific or Technical Report describes a research process or research and development results or the current state-of-the-art in a certain field of science or technology. Therefore all documents in the following list are Technical Reports, if they deal with a technical subject:

- report about laboratory experiments
- construction and design report
- report about testing and measurements
- report about internships
- various theses like project, diploma, magister, bachelor, and master thesis, doctorate thesis/dissertation, habilitation treatise
- article or report about research work in scientific journals
- project report, intermediate or final report etc.
- testing regulations
- work report
- functional specifications and functional requirements
- business plan
- expertise
- patent
- functional description, user manual, software documentation etc.

A Technical Report can be defined as follows:

▶ Technical Report = report about technical subjects, written in the "language of science and technology" (special terms and phrases, display rules etc.).

The Technical Report shall bring clarity to the reader! This means, the reader shall understand the topics described in the Technical Report in exactly the same manner as the

© Springer-Verlag GmbH Germany, part of Springer Nature 2019
H. Hering, *How to Write Technical Reports*,
https://doi.org/10.1007/978-3-662-58107-0_1

author has meant it without any feedback or answers from the author. This requirement has three aspects:

1. clarity of the technical or scientific statement (prerequisites, facts, figures, findings, conclusions, own statements)
2. clarity of what is already known and what is new as well as
3. clarity of what has been worked out by the author.

▶   Whether your Technical Report is clear enough, can be checked as follows: Imagine you are a reader who has basic technical knowledge, but no detailed knowledge about the topic or project described in the Technical Report. This fictive reader shall understand the Technical Report without any questions!

Today it is increasingly important to present your ideas and work results in Technical Reports to the scientific community, in interdisciplinary teams, to funding organizations and the interested public in a positive, professional manner. This causes sometimes trouble for engineers and natural scientists. Yet, it is not all that difficult to present one's working results in a logical, clearly reproducible and interesting way to create the impression among your audience that this work was done by an experienced professional.

It starts with taking a written report into your hands. Is it bound properly? Is it stored in a clean, tidy and wrinkle-free binder? Is there a clearly understandable title leaf? After you have got a rough overview of the contents you may ask: Does the title give sufficient and representative information about the contents of the Technical Report? If you go into more detail, the following questions may occur. Is there a table of contents? Does it list page numbers? Is the table of contents ordered by logical rules, can you recognize the "backbone"? Does the report describe the starting point of the situation or project in an understandable way? Did the author critically reflect the task at the end of the report? Does the report contain citations? Is there a list of references etc.? Can you find tables, figures and references easily and are they designed according to common rules?

This book is designed to be lying beside the PC. It shall answer questions instead of putting up new questions. You will get important information regarding how to avoid mistakes and obstacles during the presentation of your Technical Report. Moreover, this book will show you many important rules and checklists for text, table and image creation as well as for working with literature. Applying these rules and hints will make your Technical Reports readable and clearly understandable and comprehensible for your audience.

If you read this book from the first to the last page you will notice, that several information is presented more than once. This was done on purpose. Most information required to create a Technical Report is closely linked with other pieces of information. In order to present each section of this book as complete as possible in itself and to avoid too many cross-references which would disturb fluent reading, we tried to give all the information you need to complete the task which is just described in the current section of the book.

We recommend all of you who are not very experienced in writing Technical Reports to read Chap. 2 as well as Sects. 3.7 and 3.8, before writing your next Technical Report.

Each writer's problem described in this book has happened to ourselves or has occurred in Technical Reports submitted by students or during the authors supervised the writing of final study theses or doctorate theses. In addition the daily professional experience of the authors and many comments of our (German) readers have influenced the contents and layout of this book. Therefore this book reports "from practical experience for practical usage".

Planning the Technical Report

2

▶ **Preliminary thoughts prior to planning the Technical report**
Technical Reports shall be written so that they reach your readers. This requires a high level of systematic order, logic and clarity. You should keep these understandability aspects always in mind while performing the different work steps of your project to achieve a maximum of clarity for your reader. The tasks and necessary work steps are ordered by sequence into the phases *planning, creation and finishing (with check-ups)*. However, before describing the single measures in the planning process we will present a general overview of all required work steps to create a Technical Report.

2.1 General Overview of All Required Work Steps and Time Planning

The following overview shows all required work steps.

Required work steps to create Technical Reports
- Accept and analyze the task
- Check or create the title
- Design a 4-point-structure
- Design a 10-point structure
- the following worksteps are to be performed partly parallel or overlapping:

  - Search, read and cite literature
  - Note the bibliographical data of cited literature

© Springer-Verlag GmbH Germany, part of Springer Nature 2019
H. Hering, *How to Write Technical Reports*,
https://doi.org/10.1007/978-3-662-58107-0_2

- Elaborate the text (on a computer)
- Create or select figures and tables
- Develop the detailed structure

- Perform the final check
- Print copy originals or create PDF file
- Copy and bind the report
- Distribute the report to the defined recipients.

This list is complete, but the clarity can be further improved. To accomplish this, network planning is applied, Fig. 2.1.

This network plan is always repeated when the different steps to create a Technical Report are described, where the current work step is marked in gray.

Please keep in mind, that the amount of work to create a Technical Report is regularly completely underestimated. To avoid this, refer to the following tips.

▶ **Time planning:** Make a proper assumption of the required time and double the estimated timeframe! Start early enough to create your Technical Report— no later than after 1/3 of the total timeframe of your project.

▶ **Rule of thumb for the estimation of time:**
Title and structure: 1 day
Text creation: 3 pages per day
correction phase, final check, and delivery or distribution: 1–2 weeks.

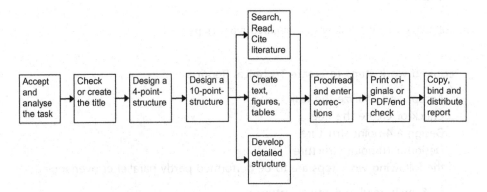

Fig. 2.1 Network plan for creating Technical Reports

2.2 Accepting and Analyzing the Task

When you write a Technical Report, there is nearly always a task, which you either selected yourself or it was defined by someone else. You should analyze this task precisely during the planning of the Technical Report.

Analysis of the task to write a Technical Report

- What is the task? Did I understand the task correctly?
- Who has defined the task?

  - a professor or an assistant (in case of a report written during your studies)
  - a supervisor
  - the development team
  - another company, who are your customer or provider
  - you yourself (e.g. if you write an article for a scientific journal)

- Which contents shall my report contain? Please write that down!
- Who belongs to the target group? For whom do I write the report? Please take notes accordingly!
- Why do I write the report? Which are the learning targets?
- How do I write the report? Which display methods and which media are used?
- Which work steps are necessary?
- Which help and assistance do I need?

  - help by people, e.g. advice-giving specialists
  - help by equipment, e.g. a color laser printer
  - help by information, e.g. scientific literature

- Does the task already contain a correct and complete title?

This work step is called "Accept and analyze the task", in the network plan it is marked in gray, Fig. 2.2.

In addition, during the planning of the report the following questions must be answered:

- Which shall be the title of the report? (develop a proposal and discuss it with the supervisor or customer)
- Which work steps that are not mentioned in the network plan need to be accomplished?
- Which background knowledge, interests and expectations do the readers of the Technical Report have?

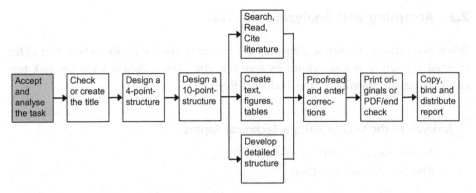

Fig. 2.2 Network plan to write Technical Reports—analysis of the task

- How do I organize the required help?
- Which help and work steps are time-critical?

2.3 Checking or Creating the Title

In the next step, Fig. 2.3, the title which in most cases is predefined by the supervisor or customer must be checked and evtl. a new title must be created.

The title of the Technical Report is the first thing a reader will notice. Therefore, it shall create curiosity to learn more about the contents of the Technical Report.

The title shall contain the main topic or the main keywords of the report, it shall be short, precise and true. It shall have a good speech melody and create interest. Explaining or additional aspects can appear in a subtitle. In any case, the title (and subtitle if applicable) shall describe the contents of the Technical Report accurately and it must not create undesired associations or wrong expectations.

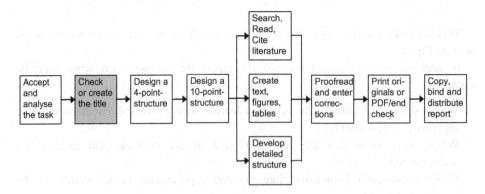

Fig. 2.3 Network plan to write Technical Reports—create the title

These demands, the title of a Technical Report must fulfil, must also be fulfilled by all other titles and headings of paragraphs, figures, tables etc.

In many cases, the task can already be used as the title of the Technical Report. Here are some examples of such tasks:

- Design of a drilling rig
- Outline of a sprayer shredding rig
- Analysis of component combinations for sales optimization
- Equipment of a meeting room with radio technology.

Even, if a title seems to be usable, we recommend that you systematically create possible title variants. Then you (and eventually the supervisor or customer) can decide which title shall be used. It is also possible to use the task as a working title in the beginning of your project. The final decision which title shall be used can be found later during your project. The following overview shows again all requirements of the title of the Technical Report as a conclusion.

Requirements of the title of the Technical Report

- The title must be clear, true, honest, short and accurate,
- it must contain the main topics or main keywords (for data base searches!),
- be short, concrete, illustrative, and believable,
- have a good speech melody,
- create interest and curiosity, i.e. create attention,
- and eventually have an additional subtitle.

Write down the main keywords, which characterize your Technical Report by hand, connect these keywords to a title, create several title variants by using different keywords and select the "best" title.

Now the process to create a title will be explained in an example.

Example for the creation of a title
We are looking for the title of a doctorate thesis. In the doctorate project a computer program has been developed, that allows the selection of the materials of designed parts depending on the stress on the part, abrasion requirements etc. The designer enters the requirements, which the material must fulfill, and the system provides the materials, which are stored in its database and match the given requirements. It has been quite early in the project that the doctorate candidate, has defined the term "CAMS" = Computer Aided Material Selection to describe the purpose of the program.

The doctorate candidate starts to create a title for his thesis as described above. He starts to manually write down the keywords that shall be contained in the title.

Title creation—Keywords
- material selection
- design
- education
- CAMS
- with computer.

Since there are many keywords, he will probably also need a subtitle to avoid that the main title becomes too long for a Technical Report. The next step is to combine the keywords to get different titles:

Title creation—Title variants
- Contribution to computer-aided material selection
- Computer-aided material selection in design
- Computer-aided material selection in design education
- Computer Aided Material Selection = CAMS
- CAMS in design education
- Help to select materials by the computer
- Computer application for material selection
- CAMS in design
- Design with CAMS
- Computer support in design education
- Material selection with the computer.

Since the doctorate candidate has defined the term, CAMS it shall definitely appear in the title of his doctorate thesis, so he decides to select the following title:

Title creation—Selection of the "best" title
Computer Aided Material Selection—CAMS in Design Education

The following overview of work steps summarizes the process to find a good title for your Technical Report.

Work steps to create the title

- write down the task
- write down the keywords which characterize the report
- combine the keywords to a title

- find new titles by varying the usage of these keywords
- read possible titles aloud to optimize the speech melody
- select the "best" title.

After the title has been created, the next step is to design the structure.

2.4 The Structure as the "Backbone" of the Technical Report

In our network plan to create Technical Reports we have now arrived at the two last work steps in the phase of planning the report. These work steps are designing the 4-point- and 10-point-structure (Fig. 2.4). Since designing the structure is the main step of planning the Technical Report, we want to introduce this important work step in the next sections.

You will get to know the relevant standards, get tips for the logic and formal design of the headings as well as concrete example structures and a little more abstract structure patterns for your orientation.

Many people do not distinguish properly between the terms "structure" and "table of contents" (ToC). Therefore, we define these terms as follows:

▶   **Structure:** *without* page numbers, contains the logic, is **intermediate result;**
         ToC: *with* page numbers, allows searching, is **final result.**

The typographic design or layout of the structure or table of contents is not a work step in the phase of planning the Technical Report, but it belongs to creating the Technical Report. Therefore, it is described in Sect. 3.1.2.

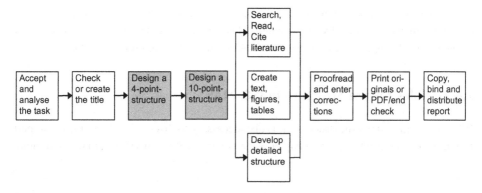

Fig. 2.4 Network plan to write Technical Reports—creating the 4-point and 10-point structure

2.4.1   General Information About Structure and Table of Contents

The structure (while writing the Technical Report) or the table of contents (after finishing the Technical Report) is the "front entrance door" into your Technical Report. It is the next piece after title leaf and Preface/Foreword and/or Summary that is read in larger documents like books, applications for research projects, final reports of research projects, design descriptions, theses, manuals etc.

▶   A good structure is so important for the understandability and plausibility of texts—even of short texts like e-mails—, that you should always structure every text that exceeds the amount of about one page with intermediate headings—at least every text describing facts.

The structure allows you to get a quick overview

– to find your way into the contents of the Technical Report,
– to get help from your supervisor, and
– to evaluate/grade your Technical Report.

▶   Therefore, you should always take the current state of the structure with you when you are going to discuss the status of your project with your supervisor (boss, assistant, professor, etc.) or with your customer.

Other materials which are not necessarily required (e.g. literature references and copies which are important or difficult to get) should also be available in the meeting.

▶   For each reader of a Technical Report the structure is the most important tool to understand the contents. Therefore, you should not make any compromises with yourself when designing the structure! This also holds true for writing the whole Technical Report. Wherever you are not confident with your report, the supervisor will criticize this not so successful part of the Technical Report in most cases—and a customer will make up his/her mind.

The information, which forms your Technical Report, will only be sorted into the drawers, which are defined by the structure. Thus, creating the structure is the creative part of the work. Writing the text is just "craftsmanship", which requires only routine.

2.4.2   Rules for the Structure in ISO 2145

When explaining the term structure, it is also necessary to discuss levels of document part headings. People use terms like chapter, subchapter, section, subsection, main item, item, clause, sub clause, paragraph, listing, etc. To refer to document parts of various levels, but these terms are not used by all people in the same sense.

If you look into the standard ISO 2145 "Documentation—Numbering of divisions and subdivisions in written text", you will find that the standard uses the terms "main divisions" for the 1st level, "subdivisions" for the 2nd level and "further levels of subdivision" for the 3rd and all lower levels.

However, this terminology does not comply with the general usage of language of most people, who think of large documents being subdivided into chapters, subchapters, sections, and subsections.

Document part headings
title (whole report)
chapter
subchapter
section
subsection.

In this book, we comply with the publication standards of the editor and use the terms chapter and section for parts of a section on any structure level. If we want to refer to all chapter and section headings, we use the term document part headings.

The following terms should be used to refer to text elements within the chapters and sections.

Text elements
paragraph
sentence
word
character
bullet list with item characters like—or •
numbered list (or ordered list) with 1., 2., 3. or (a), (b), (c).

Apart from text, the document parts can also contain other objects that illustrate the statements or messages given in the text. In many texts the following objects, which are equivalent to paragraphs occur.

Illustrating elements
table
figure
equation
list.
un-numbered list with item characters like – or •
numbered list with numbers or letters like 1., 2., 3. or a), b), c)
legend.

The standard ISO 2145 "Documentation—Numbering of divisions and subdivisions in written text" is the most important standard for creating the structure of a document. It is relevant for all types of contents, i.e. for texts dealing with technology, commerce, humanities, laws, medicine etc. and for all kinds of written documents like manuscripts, printed works, books, journal articles, manuals, directions for use and standards.

The standard itself has the following structure:

- Scope and field of application
- Numbering of divisions and subdivisions
- Citation of division and subdivision numbers in text

  - Division numbers shall not have more than three levels.
  - Points are used in section numbers only *between* the structure levels.
  - i.e. "1 Introduction" and not "1. Introduction".
  - The chapter number "0" may be used for a preface.
  - The subchapter number "n.0" may be used in a chapter, if the subchapter has the character of a prolog, foreword, preface or introduction.

The numbering of document parts is in consecutive Arabic numerals. Each document part can be further subdivided into at least two subdivisions. The subdivisions are also continuously numbered. The document part hierarchy is expressed by a full stop between the numbers of subdivisions on different levels. No full stop shall be used at the end of the final level, i.e. chapter numbers will not have a full stop at the end.

According to ISO 2145, there can be any number of document part levels, but the number should be limited, so that reference numbers are still easy to identify, to read and to cite. We recommend that the number of document part levels should be limited to three, if possible. Example: A document has nine chapters numbered 1, 2, 3, etc. Chapter two is e.g. subdivided into subchapters 2.1 and 2.2. Subchapter 2.1 is subdivided into Sects. 2.2.1.1, 2.1.2 and 2.1.3. To keep the document numbers simple, we recommend, that the number of equal document parts on the same level should not exceed nine.

Technical reports require a high level of tidiness and logic. This logic must naturally speaking be reflected in the structure. Therefore, when writing a Technical Report, the author must always keep the inner logic from the first sketch to the final version of the structure. The sequence of work steps described in Sect. 2.4.4 will nearly automatically result in a good and logical structure. Before this is described in detail, we want to introduce important rules for document part numbers and document part headings, because these rules will be applied in Sect. 2.4.4 during the creation of structures.

2.4.3 Logic and Formal Design of Document Part Headings

Document part numbers and document part headings express the logic of the sequence of thoughts and work steps (the "thread" or "backbone") in the Technical Report. For many people "logic" has something to do with mathematics and its rules. But there is also the logic of language, which is examined in many intelligence tests beside the mathematical logic.

▶   You should be able to optimize your own structures according to the logical sequence of thoughts and work steps described in your Technical Report. This requires that you develop the ability to check your own structures for proper logic of language.

This recommendation will now be explained by means of examples and further descriptions. It is a key requirement of a logical structure that different document part headings on the same level of hierarchy must be equally important and consistent. Therefore, the following part of a structure is not logical.

Part of a structure (different structure level for equivalent sections)
wrong:

3.5 Technical evaluation of concept variants
    3.5.1 Technical evaluation table
3.6 Economical evaluation table.

It happens quite frequently in Technical Reports and other larger documents or books that a document part heading is subdivided only once. However, this is not logical, because the subdivision into document parts of a lower than the current level happens, because several aspects of a super ordinated topic shall be distinguished from each other.

Therefore it is not logical, to subdivide a higher-level topic in the next lower document hierarchy level into only one document part heading. Here you should either add one or more additional document part headings of the same hierarchy level or leave the super-ordinated topic without subdivision. Here is a correct alternative for the bad example above:

Part of a structure (correction)
right:

3.5 Technical-economical evaluation of the concept variants
    3.5.1 Technical evaluation of the concept variants
    3.5.2 Economical evaluation of the concept variants
    3.5.3 Summarizing evaluation of the concept variants in the s-diagram.

This principle must be applied on all structure levels.

On each structure level there must be at least two document part headings. Otherwise, the subdivision does not make sense. Here is an example.

Unlogical part of a structure with two possible corrections
not logical:

1   Introduction
    1.1 Starting point
2   Basics of metal powder production

logically right:

1   Introduction
    1.1 Starting point
    1.2 Goals of this work
2   Basics of metal powder production

also logically right:

1   Introduction
2   Basics of metal powder production.

Each document part heading shall be complete in itself and represent the contents of the document part properly! It shall be short, clear and accurate as the title of the whole Technical Report. Document part headings that consist of one word only can often be improved. Exceptions from this rule are generally-used single words like Introduction, References, Appendices etc.

Part of a structure with too vague entries (with correction)
bad example:

3.4 Test rig
    3.4.1 Conditions
    3.4.2 Description

better example:

3.4 Test preparations
    3.4.1 Preparing the specimen
    3.4.2 Calibrating the measuring devices
    3.4.3 Building-up the test equipment.

The following overview shows a summary of the rules mentioned so far plus additional rules for document part numbers and headings.

Rules for document part numbers and headings
Rules of logic

- Full stops in section numbers define the hierarchy level in the document.
- Document part numbers0, n.0 etc. can be used for foreword/preface, introduction etc.
- Each hierarchy level consists of at least two document parts, which are
  logically of equal importance.
- The document part heading may not be the first part of the first sentence of
  the first paragraph in the appertaining text, but it must be an own and
  independent element of the Technical Report. The first sentence of the
  following text must be a complete sentence, which may pick up or repeat
  the contents of the document part heading.

Formal rules

- The declaration in lieu of an oath, task, abstract, foreword/preface and table
  of contents always get a document part heading, but no document part
  number.
- At the end of document part number and document part heading, never use
  a punctuation mark like period, colon, question mark, exclamation mark etc.
- It is unusual to formulate the document part heading as a complete sentence
  or as a main clause with one or more subclasses.
- At the end of document part headings, there is never a reference to the
  literature like "[13]".

Layout rules

- If you want to create the table of contents automatically with your
  word-processing program, use the standard format patterns or formatting
  styles resp. in the continuous text. Format chapter headings with "Heading
  1", subchapter headings with "Heading 2", section headings with "Heading 3"
  etc. You may as well change the formatting of these format patterns to
  modify the appearance of the headings in the continuous text. To modify the
  appearance of the table of contents, change the format patterns "ToC 1",
  "ToC 2" or how ever they are called in your word processor. It is usual, that
  the document part headings appear in boldface typing and larger than the
  normal text. They must not be underlined.
- Please avoid capital letters in headings and table items (in the table of
  contents, list of figures, list of tables etc.), because this is substantially more
  difficult to read than the ordinary mixture of capital and small letters.

– The meanwhile withdrawn ISO 5966 "Documentation—Presentation of scientific and Technical Reports" defined that document part headings should
be divided from previous and following text by one empty line above and
below the heading. Whereas ISO 8 "Documentation—Presentation of periodicals" defines, that the distance above document part headings should be
larger than below them. If the distance above a document part heading is
larger than the distance below, it becomes clearer, which heading belongs to
which text, and therefore we recommend this layout principle.

The rules above hold similarly true for titles of tables and figures/illustrations with the
following exceptions:

– At the end of table and figure titles, there must appear a citation, if the figure or table is
created by other authors.
– There are other rules for table numbers and figure numbers than for document part
numbers. Figures and tables are either chronologically numbered through the complete
Technical Report or the numbers are combined using the chapter number and a running number within the current chapter. Often these two components of the table or
figure number are connected by a hyphen, see Sects. 3.3.2 and 3.4.2.
– If your word processor shall automatically create the list of figures and list of tables
from the figure and table titles, you must not use manual paragraph formatting to
influence the appearance of the text, but you should apply appropriate paragraph
formatting styles or labels.

After we have introduced you to the most important rules for the formulation and
layout of document part numbers and headings, now we can use that knowledge to create
the structure.

2.4.4 Work Steps to Create a Structure and Example Structures

The creation of the structure should be divided into several consecutive work steps.
Starting from the working title (or the final title) the main topic or core message of the
Technical Report should be formulated in one sentence. This information will then be
further subdivided into document part headings up to the complete final structure, which
will appear in the Technical Report as the table of contents later. To develop the final,
logical structure, the procedure shown in the following overview has quite frequently been
successfully applied.

Work steps to create a structure

1. Formulate the title of the main topic, main target or core message of the Technical Report in one sentence.
2. Subdivision into 3–4 main items (4-point-structure).
3. Further subdivision into 8–10 main items (10-point-structure).
4. Further subdivision of extensive main items.
5. Further subdivision into the final detailed structure parallel with the further elaboration of the Technical Report.
6. Last but not least: Check whether the document part numbers and headings are identical in the structure and in the text (check for completeness and correctness) and add page numbers to the structure to make it a table of contents, if the table of contents shall not be automatically created by your word processor.

If you apply this procedure, the logical order of information, which is already defined in the 4-point-structure, cannot be lost any more up to the final detailed structure, when you add divisions or split divisions into subdivisions!

Now this procedure shall be explained by means of examples. The examples are derived from a report about the enhancement of the computer network at a customer company (a project report), a design report, a report about executed measurements (laboratory report), and a diploma thesis, where a computer program has been developed. Naturally speaking, this procedure can be applied for any other type of report like for literature research works etc.

Example 1: Design report

Title of the report: Redesign of a production plant for Magnesium-Lithium-Hydrogen alloys

1st Step: Formulate main topic (main target) of the Technical Report

Weaknesses of the existing founding plant shall be improved by the redesign

2nd Step: Subdivision into 3–4 main items (4-point-structure)

State of the art
Description of the existing weaknesses
Description of the modifications

3rd Step: Subdivision into 8–10 main items (10-point-structure)

1 Introduction
2 State of the art
3 Necessary modifications of the existing plant
4 Requirements the new plant shall fulfill
5 Redesign and reconstruction of the existing plant
6 Practical testing of the new plant
7 Evaluation of the tests with the new founding plant
8 Conclusions and outlook

4th Step: Further subdivision of extensive main items

Chapter 3 can be subdivided into the necessary modifications (possible usage of the plant for other technological processes, facilitated usage and handling, facilitated cleaning, improved safety while working with hydrogen etc.).

Chapter 5 can be subdivided into basic design principles applied for the redesign of the founding plant and design details.

5th Step: Further subdivision into the final detailed structure parallel with the further elaboration of the Technical Report

5 Redesign of the new plant
   5.1 Basic design principles and principle drawing
   5.2 Design details to realize the required modifications
      5.2.1 Basic design of the founding plant
      5.2.2 Temperature flow in the plant components
      5.2.3 Gas flow of inert gas and alloy gas
      5.2.4 Modifications of the casting device
      5.2.5 Flexible structure of the cast container via plugging system
      5.2.6 Inert gas container for the die-cast
      5.2.7 Central plant control via the control panel.

Example 2: Report about executed measurements
Title of the report: Damage detection with holographic interferometry

1st Step: Formulate main topic (main target) of the Technical Report

The deformation of a steel container under inner pressure shall be measured with holographic interferometry (target: identification of the influence of container geometry, welding zone, heat affected zone and intentionally added material flaws on the deformation of the steel container).

2nd Step: Subdivision into 3–4 main items (4-point-structure)

State of the art
Testing plant design
Test execution
Test results

3rd Step: Subdivision into 8–10 main items (10-point-structure)

1   Introduction
2   State of the art
3   Testing plant design
4   Test preparation
5   Test execution
6   Evaluation of the interferograms
7   Estimation and classification of measurement flaws
8   Proposals for continuing works
9   Conclusions

4th Step: Further subdivision of extensive main items

Chapter 5 can be subdivided by the type of the executed work steps into estimation of the required inner pressure in the testing container, description of the unintended material flaws in the welding zone, the heat affected zone and the intentionally added material flaws, description of the measurement points, influence of the container geometry.

In Chap. 6 the evaluation of the measuring results can be subdivided into the local influence of the weld seam and welding zone and the types of intentionally added material flaws.

5th Step: Further subdivision into the final detailed structure parallel with the further elaboration of the Technical Report

6   Evaluation of the interferograms
    6.1   Relative deformation extremae
    6.2   Influence of the heat affected zone
    6.3   Influence of the welding bead
    6.4   Influence of the intentionally added material flaws.

Example 3: Report about the enhancements of a computer network
Title of the report: Equipment of a meeting room with radio technology

1st Step: Formulate the main topic (main target) of the Technical Report

The computer network in the customer company shall be enhanced so that there are two additional internet access points for external staff members in the training room and two additional internet access points for training participants in the lounge.

2nd Step: Subdivision into 3–4 items (4-point-structure)

Analysis of the customer's requirements
Planning of the new network structure
Realization of the network enhancements in the customer company
Billing and payment

3rd Step: Subdivision into 8–10 items (10-point structure)

1 Introduction
2 Analysis of the customer's requirements
3 Planning of the new network structure
4 Preparing work steps
5 Realization of the network enhancements in the customer company
6 Inspection
7 Billing and payment
8 Conclusions

4th Step: Further subdivision of extensive main items

Chapter 2 can be subdivided into the steps status quo-analysis and target situation-analysis. Chapters 3, 4 and 5 and 9 Appendices have also been subdivided further in the original work.

5th Step: Further subdivision into the final detailed structure parallel with the further elaboration of the Technical Report

3 Planning of the new network structure
   3.1 Collection of offers from hardware suppliers
   3.2 Benefit analysis and decision of suppliers for the hardware to be used
   3.3 Planning the wiring
   3.4 Planning of external services.

Example 4: Report about the development of software
Title of the report: Computer-aided analysis and optimization of the understandability of technical texts

1st Step: Formulate the main topic (main target) of the Technical Report

Starting from existing approaches to improve the understandability of texts an interactive computer program shall be developed that measures the understandability of text and that stepwise improves the understandability of the text in constant dialogue with the user.

2nd Step: Subdivision into 3 to 4 main items (4-point-structure)

Approaches to measure and improve the understandability of texts
Development of the understandability improvement concept of <program name>
The program system of <program name>
Documentation of the source code of <program name>

3rd Step: Subdivision into 8–10 main items (10-point-structure)

1 Introduction
2 Approaches to measure and improve the understandability of texts
3 Development of the understandability improvement concept of < program name>
4 The program system <program name>
5 Documentation of the source code
6 The practical use of <program name>
7 Further development of <program name>
8 Conclusions and outlook

4th Step: Further subdivision of extensive main topics

Chapter 2 deals with the state-of-the-art as it is described in the literature. It has been further subdivided into scientific approaches to the research on understandability, practically oriented approaches to improve the understandability and Hamburg concept of understandability.

5th Step: Further subdivision into the final detailed structure parallel with the further elaboration of the Technical Report

4 The program system <program name>
    4.1   The menu structure of <program name>
    4.2   The sequence of feature groups in <program name>
        4.2.1   General overview of the sequence of all feature groups
        4.2.2   Sequence of the feature group Typography
        4.2.3   Sequence of the feature group Clarity
            … (further sections for Logic, Shortness, Motivators)
        4.2.8   Sequence of the feature group Orthography
    4.3   Help features for searching and classification
        4.3.1   The word classification object
        4.3.2   The dictionary object.

Rules and tips for creating the structure
1st Step: Formulate main topic (main target) of the Technical Report

Here you should formulate the target of the project, the literature research, the tests, the measurements, the design, the expert opinion or the report in general. Even if it seems hard to accomplish: Write that down in one sentence only!

2nd Step: Subdivision into 3–4 main items (4-point-structure)

Examples:

"Starting situation—Own contribution—Improvements of the situation—Summary"

"State-of-the-art—Testing rig design—Test execution—Test results—Conclusions"

If you integrate the task of your project into your Technical Report as an independent chapter, then the chapter "Task" and the various chapters about fulfilling the task (general draft, detailed design, computation of loads or testing rig design, test execution, test results etc.) are each an individual chapter.

3rd Step: Subdivision into 8–10 main items (10-point-structure)

Possible structuring principles for the 3rd, 4th and 5th step are:

- from rough (overview) to fine (details)
- by time sequence or
- by starting point conditions
- by project targets
- by possible alternatives
- by components or part groups
- from input to output
- in the direction of force (forces, moments)
- in the direction of flow (data, electric power, hydraulics, pneumatics, transported material)
- by improvement steps
- by related topics
- or in the special case depending on the task

4th Step: Further subdivision of extensive main items

Possible structuring principles have already been mentioned in the 3rd step. We recommend that *before* writing the text for a chapter, you should create a temporary structure of this chapter into subchapters. In the same way, *before* writing the text for a subchapter, you should create a temporary structure of this subchapter into sections or consciously decide that no further subdivision is necessary etc. This recommendation corresponds to the sequence of work steps to create a temporary 4-point- and 10-point-structure of the Technical Report, before you start at all with writing text, searching for literature and collecting other materials.

To reach your target group you should use common document part headings, which your reader expects to find in your Technical Report. In a report about laboratory experiments, a reader would for example expect document part headings like testing rig design, test execution and test evaluation or test results. Therefore, you should use these document part headings in your Technical Report and in your literature and material collection.

5th Step: Further subdivision into the final detailed structure parallel with the further elaboration of the Technical Report

This step needs no further explanation.

2.4.5   General Structure Patterns for Technical Reports

In the following, we show you structure patterns for often written types of Technical Reports, which have been successfully used in practice. If you use such a structure pattern, you do not need to create a 4-point- and 10-point-structure.

At first we provide a structure pattern for a rough design description in which after analyzing the sub functions and the design solutions of the sub functions several concept variants are defined. These will then be evaluated according to the VDI guidelines for design methodology VDI 2222 and 2225 (see also Sect. 3.3.2).

Structure pattern of a rough design description—several concept variants
1 Starting situation
2 Task
  2.1 Task definition
  2.2 List of requirements
3 Function analysis
  3.1 Formulating the overall function
  3.2 Subdivision into sub functions
  3.3 Morphological box
  3.4 Definition of the concept variants
  3.5 Technical evaluation of the concept variants
  3.6 Economical evaluation of the concept variants
  3.7 Selection of the most useful concept variant with the s-diagram
4 Design
  4.1 Design description
  4.2 Computation of loads

5   Summary and conclusions
6   References
A   Bill of materials
B   Manufacturer documents

If you do not add manufacturer documents, just use the appendix "A Bill of materials". If you want to add printouts or plots or photocopies of technical drawings in reduced size, you can structure the appendices as follows: A Bill of materials, B Assembly drawing, C Component drawings, D Manufacturer documents.

Bill of Materials in the Technical Report and in the set of drawings
The bill of materials is actually not a part of the report, but it belongs to the set of drawings. Since the drawings are transported in drawing rolls, it has proven to be practical, to add the bill of materials twice to the Technical Report: one copy of the bill of materials is added as an appendix of the bound Technical Report and the other is added to the set of drawings in the drawing roll. If during a presentation drawings are fixed to the walls of the meeting room, the (enlarged!) bill of materials can also be hung up at the wall.

Assembly drawing in the Technical Report and in the set of drawings
Please decide, whether you want to add a photocopy of the assembly drawing in reduced size to your Technical Report. It can either be added to an appendix (directly behind the bill of materials) or used in a text chapter (preferably in the chapter "Design description").

If the photocopy of the assembly drawing in reduced size is bound into your Technical Report, the parts can be referred to in the design description with their names and in addition with their position numbers, e.g. "Handle (23)". However, when the first part name with added position number occurs in the text of the design description, you should explain that the number is a position number and refers to the assembly drawing and the bill of materials.

Now we want to look at the structure patterns again. In the following, you will find a structure pattern for a rough design description where the most useful design solutions of the sub functions are combined to only one concept variant (see also Sect. 3.3.2).

Structure pattern of a rough design description—one concept variant
1   Introduction
2   Task
    2.1   Task definition
    2.2   List of requirements
3   Function analysis
    3.1   Formulating the overall function
    3.2   Subdivision into sub functions
    3.3   Morphological box
    3.4   Verbal evaluation of the design alternatives for the sub functions
    3.5   Description of the concept variant

4   Design
    4.1   Design description
    4.2   Computation of loads
5   Summary and conclusions
6   References
A   Bill of materials
B   Manufacturer documents.

Now we want to give you a structure pattern for projects dealing with laboratory experiments or other experimental works. First, there is an important rule:

▶   Laboratory experiments must always be documented "reproducible"! Provide
     so much information, that someone else will measure the same values or find
     out the same test results, if he/she executes the experiments exactly under the
     described conditions.

Therefore, the following information may never be left out:

–   testing machine, device, or rig with manufacturer, type number and/or name, inventory number etc.
–   all parameters set or selected at the machine, device or rig
–   all measuring instruments, always with manufacturer, type number and/or name, inventory number, set or selected parameters etc.
–   tested specimens with all required data according to the appertaining standard, regulation or guideline (ISO, EN, DIN or other), taken samples
–   in experiments which are not standardized similar data regarding specimen shape, experiment parameters, temperatures, physical/chemical properties etc.
–   all measured values or test results with all parameters
–   used evaluation formulas with complete bibliographical data of used references.

Structure pattern of an experimental work
1   Target and scope of the test
2   Theoretical basics
3   The laboratory experiment/test
    3.1   Testing rig design
        3.1.1   Testing machine, plant, rig or device
        3.1.2   Used measuring instruments
    3.2   Test preparations
        3.2.1   Specimen preparations
        3.2.2   Setup of the starting conditions

3.3 Test execution
    3.3.1 Execution of the preparation tests
    3.3.2 Execution of the main tests
3.4 Test results
3.5 Test evaluation
3.6 Estimation of measurement flaws
4 Critical discussion of the laboratory experiments/tests
5 Conclusions
6 References
A Measurement protocols of the preparation tests
B Measurement protocols of the main tests.

The next example structure pattern is for manuals and instructions for the usage of complex technical products. These documents should be written by technical writers. Manuals and instructions for use are structured according to different schemes. They can be subdivided and numbered according to ISO 2145, but they can as well have document part headings without document part numbers. To provide more uniformity here, EN 82079 "Creation of instructions; Structure, contents and presentation" has been published. Among other information this standard describes, which information shall be given in which sequence in instruction manuals. Other definitions used in the following structure pattern are derived from DIN 31051 "Grundlagen der Instandhaltung (Basics of maintenance)", and VDI guideline 4500 "Technische Dokumentation (Technical documentation)". The text in manuals shall be understandable for technical laymen. The vendor of the products bears the risk of product liability.

The following structure pattern for manuals and instructions for use differs from the other structure patterns, because the individual document part headings are partially not as detailed as in the other structure patterns. This was done intentionally, because the described technical products can have very different levels of complexity and very different philosophies of use. Therefore, look at this last structure pattern only as an orientation and adopt it to your described technical product. The information can either be presented according to the structure and logic of the product (product-oriented) or according to the sequence and logic of work steps during product usage (task-oriented).

Structure pattern for manuals and instructions for use

1 Before operating the machine/device
  1.1 Important information about the machine/device
      (Definition/description of the machine/device, description of the benefits, safety notes and warnings, overview of the functions)
  1.2 Supplied/delivered scope and optional parts

1.3   Usage of the machine/device (Rules and regulations,
      safety notes and warnings, intended usage, unintended usage,
      documentation provided by third parties)
1.4   Transportation of the machine/device
1.5   Requirements regarding the site
1.6   Unwrapping, assembling, mounting and setup of the machine/device
1.7   Connection of the machine/device to supply and disposal networks
      (water, electricity, computer network etc.) and operation test
2   Operation and usage of the machine/device
2.1   Initiation of the machine/device
2.2   Functions of the machine/device during normal operation, safety notes
      and warnings
2.3   Refilling consumptive materials
2.4   Cleaning the machine/device
2.5   Preventive maintenance (maintenance, inspections)
2.6   Disposal of supporting and operating materials
2.7   Shutting-down the machine/device
3   After operating the machine/device
3.1   Finding the cause of disorder and resolving it
3.2   Ordering spare parts, wear and tear parts and electric plans
3.3   Disassembling the machine/device
3.4   Disposal and recycling of the machine/device (what? where? how?)
4   Appendices
4.1   Possible causes of disorder/Trouble shooting (what shall I do, if ...?)
4.2   Spare parts, additional parts (exceeding the supplied/delivered scope)
4.3   Glossary
4.4   Index.

Naturally speaking all structure patterns described in this section can be adopted to the described project, product, topic or task. If the supervisor has published an own structure pattern, it should be used. On the other hand, if you use the structure patterns presented in this book, you will establish a correct logical sequence of thoughts, topics, work steps etc.

2.5   Project Notebook (Jotter)

In a guideline how to write Technical Reports by Thomas Hirschberg, a professor at the University of Applied Sciences in Hannover I found the following advice.

▶   You should structure the contents of your report as early as possible and note
    all open problems, decisions and remaining work steps regarding your project
    "online" in a project notebook (jotter). Do *not* start with writing your Technical
    Report after all practical work has been completed.

Since you need your project notebook both at your workplace/in the laboratory and at
home, you should use a small booklet in DIN A5 or DIN A6 format.

2.6   The Style Guide Advances Consistency in Wording and Design

A style guide is only required for the creation of larger written documents. It is similar
with the documentation manual of a documentation or translation service provider. It has
the purpose that within a larger document the same things or ideas have the same names
(formulation of phrases, terminology) or that they are displayed in the same way (design,
layout), so that the document (or Technical Report) is consistent in itself.

Therefore in the style-guide you can collect your preferred spelling of words, wording
of phrases, special terms as well as layout rules. Let us refer to the first and fourth line of
the current paragraph. You have probably noticed that the spelling of the special term
style-guide differs from the spelling in the previous paragraph and in the heading.
However, this should not happen within the same report. Therefore, such rules are listed in
a style guide (which is written correctly again here). During the proofreading and final
check you can correct violations of the rules in the style guide with the function "Find" or
"Find and replace" of your word processor. In this way, the style guide helps to keep
consistency in formulation and design within the Technical Report.

In the following overview you can see examples of what can be listed or standardized
in a style guide. The checklist shows part of the checklist for this book (the German
version). The spelling and layout rules have been defined by the editors and the author of
this book, here the editors have mainly defined the margins and the formatting patterns
(corporate design). These formatting rules result in a packed layout. For normal Technical
Reports the font size of the standard text should be 11 or 12 pt. and the gap between
paragraphs 6 pt. All other formats listed in the checklist must be adjusted accordingly.
Similar layout rules as we have got from the editors can be found in most institutes and
companies.

It is strongly recommended, that you create and use an own style guide for your
Technical Reports. It is too time-consuming and insecure, to try to keep in mind all
defined rules and regulations.

Example entries in a style guide for a Technical Report
Bullet lists and ordered lists always start at the left document margin.

- the first list mark is a small thick bullet (Alt + 0149 while Num Lock key is pressed)

  – the 2nd list mark is a dash (Ctrl + Minus in the numeral keyboard or Alt + 0150)

Tipps and instructions (you-form, the hand symbol is not italic, it belongs to the font Wingdings and is entered via Alt-0070, indentation hanging 0.5 cm):
☞ *Write down* ...

Spelling use	do not use
design report	design-report
style guide	design-report
PowerPoint	Powerpoint

Tabs/spaces
use a tab behind a figure/table number
use a tab behind a section number
use a tab behind a chapter number

Self-defined formatting patterns

Element	Formatvorlage	Stichworte
above list	ListAbove	2 pt empty line
list	ListNextLine	10 pt, indentation hanging 0.5 cm, after 0 pt
below list	ListBelow	6 pt empty line
figure heading	HeadingFigure	9 pt/distance before 24 pt/after 12 pt
formula	Formula	10 pt, indentation 1 cm
handwriting	Handwriting	Segoe Script 9 pt/bold/indentation 0.5 cm
standard paragraphs	Standard	10 pt/distance after 4 pt
table heading	HeadingTable	9 pt/distance before 12 pt/after 12 pt double line ¾ pt
document part	Heading 1	16 pt/bold/distance after 30 pt
headings	Heading 2	13 pt/bold/distance before 18 pt/after 9 pt
level 1–3	Heading 3	11 pt/bold/distance before 12 pt/after 4 pt

Enter special characters via numeral keyboard (switch it on with FN + NumLock or FN + F11)

Standard font	Standard font	Symbol font	Symbol font
© Alt-Strg-C, Alt-0169	± Alt-0177	• Alt-0183	⇒ Alt-0222
® Alt-Strg-R, Alt-0174	– Alt-0150	≈ Alt-0187	× Alt-0180
~ Alt-0126	• Alt-0149	≅ Alt-0064	≤ Alt-0163
« Alt-0171	· Alt-0183	≠ Alt-0185	≥ Alt-0179
» Alt-0187		≡ Alt-0186	™ Alt-0212

soft line break: Ctrl-Minus dash: Ctrl-Minus in the numeral keyboard

Conclusion

This completes the planning of your Technical Report. Now the most extensive part of the work steps on your Technical Report will be described, i.e. the practical realization of your plans. This contains literature research and reading, writing text, creating figures and tables as well as the continuous adoption of your structure to the current state of your project and development of your Technical Report.

2.7 References in This Chapter

DIN 1421:1983-01 Gliederung und Benummerung in Texten
DIN 31051:2012-09 Grundlagen der Instandhaltung (Basics of maintenance)
DIN EN 82079:2018-05 Erstellen von Anleitungen; Gliederung, Inhalt und Darstellung (Preparation of Instructions – Structuring, Content and Presentation)
ISO 2145:1978-12 Documentation – Numbering of divisions and subdivisions in written documents.
VDI 2222-2225 Konstruktionsmethodik, mehrere Blätter (design methodology, several sheets)
VDI 4500 Technische Dokumentation – Benutzerinformation (Technical documentation, several sheets)
Brändle, M. et al. Praxisleitfaden Betriebsanleitungen. tekom (Hrsg.), Stuttgart, 4. Aufl. 2014
Gabriel, C.-H. et al. Richtlinie zur Erstellung von Sicherheitshinweisen in Betriebsanleitungen. tekom (Hrsg.), Stuttgart, 2005
Reichert, G. W. Kompendium für Technische Dokumentationen. 2. Aufl. Leinfelden-Echterdingen: Konradin, 1993
Erdmann, E. et al. Praxisleitfäden: Regelbasiertes Schreiben - Englisch für deutschsprachige Autoren. tekom (Hrsg.), Stuttgart, 2. Auflage 2017

Writing and Creating the Technical Report

3

▶ **Preliminary thoughts prior to the practical work on your Technical Report**
In this chapter you will get many tips and see many examples for the appropriate creation of the Technical Report. Hints for working with word processor systems are mainly collected in Sect. 3.7. However, before showing the details of this chapter, we want to present some general and summarizing thoughts.

We have already discussed that creating the structure of the Technical Report is the difficult and creative part of the whole task. The structure determines, whether the Technical Report has a comprehensible inner logic.

Despite the fact, that many beginners in "Technical Writing" find it difficult, creating the complete Technical Report is more or less a craft. It includes keeping the rules, which are introduced and explained in detail in this book.

Often you can see it in Technical Reports from when on the time-pressure in the project has raised, i.e. from when on errors and un-precise descriptions show up more frequently! Therefore, the final check is a very important work step, which may not be left out due to time-pressure at all, see Sect. 3.8.3.

It may happen that a supervisor or customer does not have contents-related reasons to criticize your work. If he/she still wants to find something, he/she often criticizes little odd details or formal aspects. To avoid these problems you should apply the tools report checklist and style guide—both tools have proven to be useful in many projects in practice.

Often institutes, companies, authorities, and other institutions have rules for the optical appearance and computer-based creation of letters, reports, overhead slides, and other

© Springer-Verlag GmbH Germany, part of Springer Nature 2019
H. Hering, *How to Write Technical Reports*,
https://doi.org/10.1007/978-3-662-58107-0_3

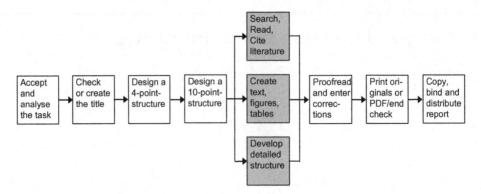

Fig. 3.1 Network plan to write Technical Reports—search and cite literature, create text, figures and tables, develop a detailed structure

documents, so that they fit into the unique optical appearance (corporate design) of the institution. Such rules and guidelines should be integrated into the style guide (Sect. 2.6) and the report checklist (Sect. 3.8.1) and applied during the whole process of creating the Technical Report.

After finishing the phase planning the Technical Report, we will now go into the details of writing and creating the Technical Report. In the network plan, this phase is marked in gray again.

Please keep in mind the following rule for all tasks marked in the network plan, Fig. 3.1:

▶   From time to time, you should imagine to be the reader and ask yourself: When do I need which information? Does the current figure appear "out of the blue"? Should I pick up the structure, write an intermediate summary, or announce the new document part from a very general point of view? Is the subdivision of information logical and comprehensible?

Prior to describing the single steps to create the report in detail, we will provide you with an overview of the general structure of a Technical Report with all parts that need to be written.

3.1 Parts of the Technical Report and Their Layout

The names, contents and order of the parts of a Technical Report in general are defined in ISO 7144 "Documentation – Presentation of theses and similar documents". It is shown in the following list.

Parts of a Technical Report or a thesis according to ISO 7144

front matter

- outside and inside front cover (cover pages 1 and 2)
- title leaf
- errata page(s)
- abstract
- preface
- table of contents
- list of illustrations (figures) and list of tables
- list of abbreviations and symbols
- glossary

body of thesis

- main text with essential figures, illustrations and tables, list of references

annexes

- tables, figures, illustrations, bibliography etc.

end matter

- index(es)
- curriculum vitae of the author
- inside and outside back cover (cover pages 3 and 4)
- accompanying material

In Germany, this is standardized in DIN 1422. The order of the required parts of the Technical Report as defined in DIN 1422 is slightly different from the order defined in ISO 7144.

Not all parts are necessary or required in all Technical Reports. It is the writer's duty to ask the supervisor or customer which rules and guidelines must be kept as long as they are not available in written form.

In the following, we will introduce the individual parts of the Technical Report and give some hints regarding their layout.

3.1.1   Front Cover Sheet and Title Leaf

After the "best" title has been developed in Sect. 2.3, the absolute and relative position of all parts that must appear on a front cover sheet and/or title leaf must be defined. A front cover sheet and/or title leaf is a must for a Technical Report.

You should distinguish the (inner) title leaf from the (outer) front cover sheet. The front cover sheet is the title visible when the Technical Report lies on a table as a closed book.

The title leaf is only visible after you opened the Technical Report and in most cases after you turned a blank white sheet of paper.

However, if the Technical Report is bound so that the (outer) front cover sheet is a transparent sheet of plastics, then inner and outer title are identical, i.e. the information provided on the (inner) title leaf and the (outer) front cover sheet are identical. Then the blank white sheet of paper which usually follows the front cover sheet will follow the title leaf instead.

Beside this special case, the following rule holds true: the title leaf always contains more information than the front cover sheet. For instance, in Technical Reports written during study courses it is unusual to list the supervisors on the front cover sheet, whereas they definitely have to be listed on the title leaf.

There are some faults which occur quite frequently on front cover sheets. Some of them are displayed in Fig. 3.2.

The faults occurring most frequently on front cover sheets are:

–   The name of the institution is missing on the top of the page.
–   The name of the university is correctly specified, but the name of the department and/or institute are missing.
–   The title (essential!) is layouted with a too small font size, while the type of report (not so important!) is much larger than the title.

The layout of the front cover and title leaf are also influenced by rules that are more general. For example, the corporate design of a company or university may define that for a special type of Technical Report a specific form, e.g. "Cover for laboratory reports"

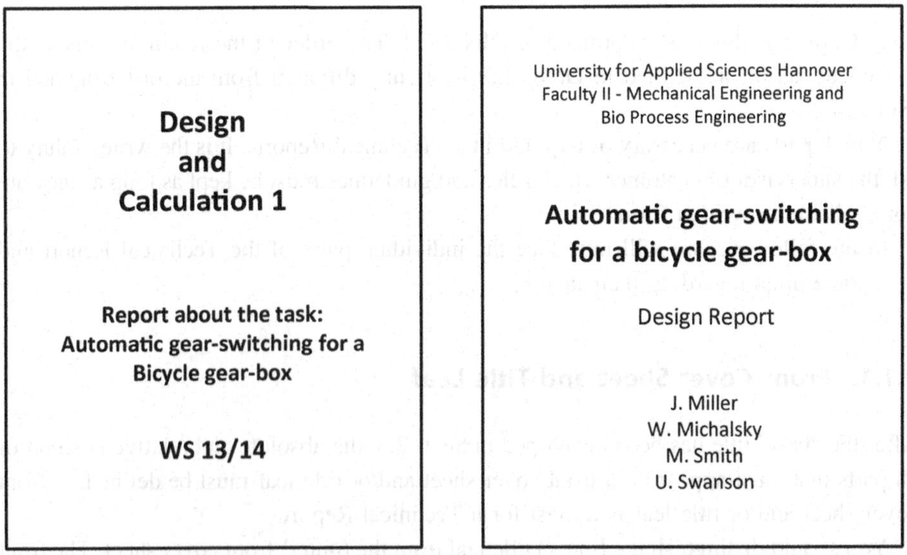

**Design
and
Calculation 1**

**Report about the task:
Automatic gear-switching for a
Bicycle gear-box**

WS 13/14

University for Applied Sciences Hannover
Faculty II - Mechanical Engineering and
Bio Process Engineering

**Automatic gear-switching
for a bicycle gear-box**

Design Report

J. Miller
W. Michalsky
M. Smith
U. Swanson

Fig. 3.2 Comparison of a faulty (left side) and a correct (right side) front cover sheet for a design report

must be used. In most cases there are also rules for the layout, e.g. that the company logo must always be located at the top and on the right or left side or in the middle. Other layout rules define which font type and font sizes must be used. In universities, these rules may exist for an institute, a department, or for the whole university. Naturally speaking, you should follow these rules.

The right version in Fig. 3.2 is fine as long as it does not disobey existing rules of the university or customer. Now we want to look at the steps how to develop the front cover and the title leaf of your Technical Report. The example to explain the work steps is again the dissertation "Computer-aided material selection – CAMS in design education".

Our doctorate candidate has got the information from his university which rules should be followed when designing a front cover and title leaf, and he has looked at other dissertations in the university library to see good examples for the application of these rules. The following information must occur on the title leaf of a dissertation:

- the title of the work,
- the additional specification "Dissertation to qualify for a doctorate degree at the University of Klagenfurt",
- the names of both supervisors and the author with full academic titles as well as
- the city, where the university is located, together with month and year when the dissertation is submitted.

There are no exact rules for the positioning of the information on the title leaf. Therefore, the doctorate candidate has manually written down four variants how his title leaf could look like, to judge the line breaking of the single information blocks and to get an impression of

- the proportions of text and intermediate white space,
- the justification of the text blocks (centered, left justified, right justified, or along a line) and
- the line breaks,

see Figs. 3.3 and 3.4.

The doctorate candidate then types his favorite version into his word processor. There the following typographic design options are optimized:

- font type and font size and
- methods to emphasize text like bold, italic, expanded spacing etc.

The title leaves for reports and theses written during a study course like diploma thesis, project reports, laboratory, and design reports contain slightly different information from the information in this example.

In case of a diploma thesis the student ID number and start and end date of the diploma project must be specified. In case of project reports or other university-internal reports the student ID number and the term or semester are specified. Please refer to the front cover and the title leaf of a diploma thesis, Fig. 3.5, and a design report, Fig. 3.6.

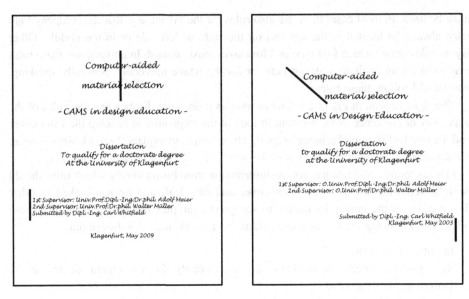

Fig. 3.3 Two handwritten drafts of the title leaf of a dissertation (the placement of information varies between centered, left justified, along a line, and right justified)

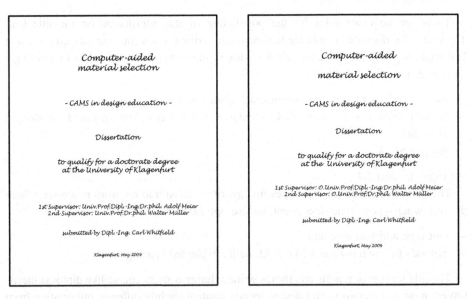

Fig. 3.4 Two handwritten drafts of the title leaf of a dissertation (with variation of font size and line breaks)

Fachbereich
Maschinenbau

Hochschule
Hannover

Design of a test plant
to examine the static strength
of valves

Bachelor thesis

Bart Wayne
Thomas Miller

Fachbereich
Maschinenbau

Hochschule
Hannover

Konstruktion einer Vorrichtung
zur Dauerfestigkeitsprüfung
von Ventilen

Bachelor thesis

submitted by: Bart Wayne, 945672
Thomas Miller, 942388

1st Supervisor: Prof. Dr. L. Heinrich
2nd Supervisor: Prof. Dr. F. Grambach

Start: 20.3.2014
End: 20.6.2014

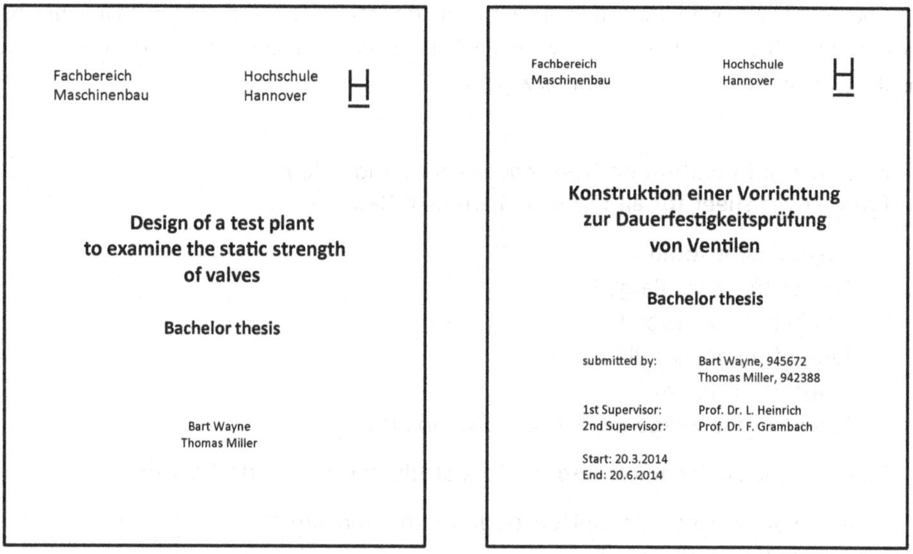

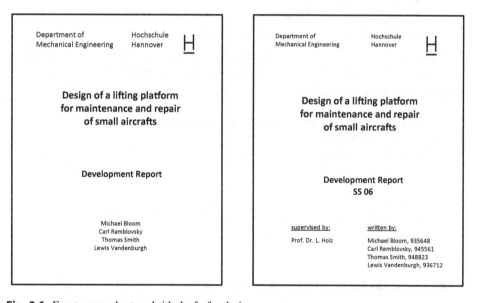

Fig. 3.5 Front cover sheet and title leaf of a diploma thesis

Department of
Mechanical Engineering

Hochschule
Hannover

Design of a lifting platform
for maintenance and repair
of small aircrafts

Development Report

Michael Bloom
Carl Ramblovsky
Thomas Smith
Lewis Vandenburgh

Department of
Mechanical Engineering

Hochschule
Hannover

Design of a lifting platform
for maintenance and repair
of small aircrafts

Development Report
SS 06

supervised by:
Prof. Dr. L. Holz

written by:
Michael Bloom, 935648
Carl Remblovsky, 945561
Thomas Smith, 948823
Lewis Vandenburgh, 936712

Fig. 3.6 Front cover sheet and title leaf of a design report

The following overview summarizes again the minimum information that must be provided ("What") and their location on the front cover sheet and title leaf together with a qualitative specification of the font size ("How").

Minimum information on front cover sheet and title leaf
<u>Front cover sheet</u> for all types of Technical Reports:

  (Logo and) institution
  Title of the work (large!)
  Subtitle (if applicable)
  Type of report (smaller!)
  Author/s (medium)
  Characteristic image or illustration (if applicable)

Title leaf for all Technical Reports in a study course beside final theses

  (Logo and) institution: university/department/institute
  Title of the work (large!)
  Subtitle (if applicable)
  Type of report (smaller!)
  in the subject <name of the subject>
  Specification of semester or term (e.g. SS 09)

  supervised by: written by: name/s or group and group number
  (name with title/s)   (first name/s, name/s, student ID number/s)

Title leaves for final theses

  (Logo and) institution: university/department/institute
  Title of the work (large!)
  Subtitle (if applicable)
  Type of report (smaller!)

  1st <u>Supervisor</u>: (name with title/s)   written by:
  2nd <u>Supervisor</u>: (name with title/s)   (first name/s, name/s, student ID number/s)

  Start: (exact date)
  End: (exact date)

Title leaves for Technical Reports in industry

  (Logo and) company, main department, department
  Title of the work (large!)
  Type of report (smaller!)
  written by

> Author/s (title/s, first name/s, name/s, department/s, eventually e-mail, telephone, fax, eventually addresses of contact persons, promoters, sponsors etc.)
> Date and eventually version (e.g. June 2014 or Version 1, June 2014)

This completes the description of which information is placed where in which layout on the front cover sheet and the title leaf. The following overview summarizes again the work steps to design the front cover sheet and title leaf.

> **Placement of information on front cover sheet and title leaf**
> Work steps to place the information on a front cover sheet and title leaf:
>
> - create several variants, use handwriting on paper to avoid restricting your creativity by a limited screen
> - try out different line breaks
> - form different blocks of information (title, supervisors, company/university, date)
> - arrange these blocks centered, left justified, right justified or along an angular line
> - select the "best" arrangement
> - transfer it to your word processor and optimize it there
> - care for layout rules of your university, institute, or company

More details about what information has to be placed on front cover sheet and title leaf can be found in ISO 1086 "Information and Documentation - Title leaves of books" and in ISO 7144 "Documentation - Presentation of theses and similar documents".

After the front cover sheet and title leaf the task, declaration in lieu of an oath (for bachelor, master and diploma theses etc.), acknowledgements, and preface may occur. The next part of the Technical Report is always the table of contents.

3.1.2 Structure with Page Numbers = Table of Contents (ToC)

In Sect. 2.4 "The structure as the 'backbone' of the Technical Report" it has been stated, that the document part headings in the structure contain the inner logic of the Technical Report. The structure defines the sequence and the logical super- and subordination of the document part headings. However, in that shape it is not yet suited to look up and search specific passages in the text. Only after adding page numbers, the structure becomes a table of contents and then it is suited to look up text passages. Therefore the table of contents of a Technical Report must *always* have page numbers for all document part headings from level 1 to 3 (4).

In this section, we will deal with the layout and formal design of the table of contents on the paper. By the way, the table of contents of this book is a good guideline for your tables of contents.

The headline of the table of contents is *not*—as frequently written—"table of contents". The headline is just "Contents". The fact that it is a "table of …" becomes clear at first sight on any page of the table of contents. Now we want to present some thoughts regarding page numbers and page numbering.

The page numbering always begins on the first text page. This is the page, which displays the chapter number "1". The chapter number "0" may occur, if this chapter is a foreword, a preface, an introduction, or other division of similar type, see ISO 2145. In front of the first text page there may occur other pages, especially the table of contents. These parts of the Technical Report are called *front matter*.

Whether the parts of the front matter occur in the table of contents at all and whether they occur with or without page numbers is treated very different. Nearly every book uses different rules for this problem. Therefore, we want to make a proposal, how you can solve this issue in your Technical Reports.

– The front matter can get Roman page numbers or no page numbers at all. If the front matter gets page numbers, the title leaf is the first page of the front matter. It is integrated into the page numbering with Roman page numbers, but it does not get a page number printed onto the page. If you apply **book page numbering** with page numbers on the front and reverse side of the pages, the reverse side of the title leaf also does not get a page number. Therefore, the printed page numbers start with III on the first page of the foreword/preface or table of contents.

– If you apply the common **report page numbering** with page numbers only on the front side of the pages, the first page of the foreword/preface or of the table of contents, which follows the title leaf, will get the page number II. The rest of the table of contents and the other parts for the front matter will also get Roman page numbers. However, in small and medium-sized Technical Reports the front matter should not get page numbering.

– The table of contents (ToC) shall list the parts of the front matter in the right order but without page numbers, so that in the table of contents there are only Arabic page numbers. The Roman page numbers of the front matter are much wider than the Arabic page numbers of the normal text chapters. The table of contents should also list the parts of the Technical Report which occur behind the indexes and appendices, like "Curriculum vitae" and "Declaration in lieu of an oath" in dissertations in the right order, but without page numbers.

ISO 7144 establishes the following rules regarding page numbering:

– Title leaves are integrated into the page numbering, but they do not get a page number.
– The pages are consecutively numbered with Arabic numbers, empty pages and the front matter are counted as well. The first page number occurs on the front side of the first printed page.
– The pages of the annexes/appendices will get own page numbers with Arabic numbers which contain the letter of the annex/appendix and the page number starting from 1.

In the table of contents, the page number being listed is only the first page number of any document part. A frequently occurring mistake in Technical Reports is to list start and end page number of the document parts with an extension mark in between.

wrong   5.1 Experiment set-up 35–36
correct   5.1 Experiment set-up 35

The page numbers in the table of contents are always printed right justified.

Now the placement of the document part headings still needs to be discussed. ISO 2145 "Documentation – Numbering of divisions and subdivisions in written documents" provides a layout example for a table of contents where independent of the hierarchy level in the document all document part numbers are aligned along a common building line. All document part headings are aligned along another common building line more to the right. Indentations are not recommended in ISO 2145. This kind of layout is shown in the following example.

Table of contents of a chapter according to ISO 2145

5	Welding experiments with the optimized extraction system	34
5.1	Experiment Set-up	35
5.2	Preparations for the experiments	36
5.2.1	Definition of the experiment program	38
5.2.2	Used equipment	41
5.2.3	Used materials and expendables	45
5.3	Experiment execution	47
5.3.1	Program flow chart for the experiment execution	50
5.3.2	Additional remarks regarding the experiment execution	51
5.4	Experiment evaluation	52
5.4.1	Experiment evaluation based on the weld seam quality	53
5.4.2	Experiment evaluation based on the suction system effectivity	57
5.4.3	Assessment of handling and visibility conditions	59
5.4.4	Evaluation of flaws	60

However, since many decades tables of contents are often layouted with indentations, as it is shown in the following example.

Table of contents of a chapter with indentations for a better overview

5   Welding experiments with the optimized extraction system 34
    5.1   Experiment set-up 35
    5.2   Preparations for the experiments 36
          5.2.1   Experiment program and used equipment 38
          5.2.2   Used materials and expendables 45
    5.3   Experiment execution 47
          5.3.1   Program flow chart for the experiment execution 50
          5.3.2   Additional remarks regarding the experiment execution 51
    5.4   Experiment evaluation 52
          5.4.1   Experiment evaluation based on the weld seam quality 53
          5.4.2   Experiment evaluation based on the extraction system effectivity 57
          5.4.3   Assessment of handling and visibility conditions 59
          5.4.4   Evaluation of flaws 60

A structure or a table of contents with indentations is much clearer and is therefore recommended!

To achieve this result you should use indentations and tabs. If you use space characters, it is not possible to keep the vertical building lines precisely. If each level in the document hierarchy starts at an own building line, the reader can comprehend the inner structure of the Technical Report much better. In addition, the author can constantly check the logic of the report when writing the 4- and 10-point structure and the detailed structure.

Next, it will be shown how much the checking of the logical order of document part headings is facilitated by the indentations. Read the document part headings on the second level along their building line. You can read the following terms: "Experiment set-up", "Preparations for the experiments", "Experiment execution" and "Experiment evaluation".

A check of the inner logic ("backbone") results in the following thoughts. After the description of the experimental equipment there is a description of the preparations, which have to be executed before the experiments can be started. Then there is a description of how the experiments are executed and an evaluation of the measured results. Conclusion: the inner logic is properly built up!

These constant checks for the logic of the report during writing are effectively supported by the indentations in the structure and the table of contents.

Beside by means of indentations the structure of the document part headings can also be optically emphasized by boldface typing or larger font size, e.g.

- chapters 14 pt (bold),
- subchapters 12 pt (standard) and
- sections 11 pt (standard).

Document part headings should never be printed in upper-case letters only (capital letters or small capital letters), because the eye is not used to it. Thus, the headings are much harder to read. The reason is, that during reading eye and brain process the word contours like a picture as a "skyline" and compare them with formerly stored "skylines". If capital letters are used, the letters have to be read one after the other and the meanings of the words have to be analyzed.

Moreover, groups of document part headings can be built up by a variation of the vertical distance between the document part headings in the table of contents. Leading characters (dots) should be used between document part heading and appertaining page number to facilitate reading. The following example shows, how these mechanisms can interact effectively.

If a document part heading does not fit on one line any more (because it is too long or indented), it must be continued in the next line/s at the appropriate building line for the text of the document part headings of the relevant hierarchy level. The leading characters (dots) and the page number appear in the last line of this document part heading. The page numbers should be formatted with the same font size without any accentuations.

It is clearer and more pretty, if you insert a space between the document part heading in the table of contents and the tab with the leading dots as well as between the tab and the page number. You can search for the tabs with Edit—Search... Search for "^t" and replace the tabs with space, tab, space.

Typographic accentuations make the structure of the table of contents even clearer

Contents

1   **Introduction** 1
2   **System Requirements** 3
    2.1   Hardware 3
    2.2   Software 3
3   **Installation** 4
    3.1   Files on the installation disk 4
    3.2   Command sequence for the installation 5
4   **Program Usage** 6
    4.1   Submenu „File" 6
        4.1.1   File—New 7
        4.1.2   File—Open 7
        4.1.3   File (or Window)—Close 8
        4.1.4   File—Save 8
        4.1.5   File—Save as 8
        4.1.6   File—Print 9
        4.1.7   File—Exit 9
    4.2   Submenu „Edit" 10
        4.2.1   Edit—Undo 10
        4.2.2   Edit—Cut 11
        4.2.3   Edit—Copy 12

    ...

The gap between all document part headings and the appertaining page number should be filled with the same leading dots. The tab with the leading dots should not be formatted in bold face typing. This is the same for chapter headings. The font size should also be the same in the gaps, because dots have a larger distance and diameter at 14 pt font size compared with 10 pt, as shown in the next example.

Avoid leading dots (and page numbers) in bold type and different font size

...

3.2   Command sequence for the installation 5
4   Program usage **6**
    4.1   Submenu „File" **6**
        4.1.1   File—New 7
        4.1.2   File—Open 7

...

Subheadings without an own document part number do never occur in the table of contents. Examples are subheadings marked with "a), b), c)",α), β), γ)", or just with bold type. Unnumbered subheadings are used, e.g. to avoid a subdivision into the fourth level. They can also occur within calculations.

Different from document part headings, subheadings may have a colon at the end, if they have an announcing character.

If there are four or more hierarchy levels, the indentations reach relatively far to the right. Therefore, many document part headings can stretch across more than one line. In addition, the vertical distance to the superordinated document part heading becomes quite large. To avoid these problems there are several options:

– The table of contents may be layouted in a smaller font size than the normal text, e.g. normal text in 12 pt and the table of contents in (10 or) 11 pt.
– Subheadings without document part numbers can be used (see above).
– Subheadings with document part numbers are used, which do not occur in the table of contents (actually this should not happen!) in Technical Reports, see next item.
– Subheadings with document part numbers are used in the text, which do not occur in the overall table of contents. However, they are listed in a detailed capitular table of contents. This approach is sometimes applied in larger documents (e.g. in manuals and textbooks), but it is rather unusual for Technical Reports.
– The indentations are smaller than the widest document part number. This causes less document part headings to stretch across more than one line.

▶ If such problems occur in the table of contents, the author should discuss with
his supervisor or customer, which variant is to be used. In doubt, the classical
solution with an overall table of contents in the front of the Technical Report
showing all numbered document part headings should be preferred.

Since the options to format an automatically created table of contents after its creation
are not so well-known, in practice the tables of contents are either created with the default
layout settings, which looks ugly and is not very clearly arranged, or typed as normal text
and layouted accordingly. If they are manually typed, mistakes can creep in. You should
especially look at these points:

- All document part numbers and document part headings in the table of contents must
be exactly consistent with those in the text. Each document part heading gets its own
(begin) page number in the table of contents.
- During the end check prior to the final printout you should check, whether all numbered document part headings are listed in the table of contents with the right document part number, the right wording, and the right page number!
- If you let the computer create the table of contents, you should update and print it once
again now.

3.1.3 Text with Figures, Tables, and Literature Citations

The "text" contains all information presented in the chapters (e.g. starting with Introduction and ending with Summary). In the text, you will also find tables and figures
eventually with legends, formulas, literature citations and footnotes. Information, which is
not written by the author but cited as a base for the author's ideas or argumentations, must
be clearly marked as a citation, see Sect. 3.5. This is also necessary for cited figures and
tables.

The individual parts of the Technical Report like tables, figures, literature citations,
text, and formulas are described in more detail in separate sections within this chapter.
Here in the beginning of the chapter some aspects shall be discussed, that cannot be
assigned to the more detailed subchapters because of their general relevance.

The author of a Technical Report shall guide the reader with his words. All intermediate thoughts, conclusions, etc. shall be explicitly communicated to the reader in the text.
Thus, the reader can follow all logical thoughts of the author regarding the sequence of the
report and follow the thread in his own mind. This improves the understandability of the
Technical Report very much.

However, if a figure, a table or a bullet list follows directly upon a document part
heading, the reader is often somewhat left alone. In most cases an "introductory sentence"
is missing here, which announces and explains the logic thread or the contents of the
figure, table, or bullet list.

This deficiency occurs so often, that the author uses the abbreviation "iS>" at the left margin. The peak of the angle points to the gap between document part heading and figure, table, or bullet list and the abbreviation denotes: the "introductory sentence" is missing here. Look at the following example:

Introductory sentence between document part heading and table

... <Text> ...

3.3 Morphological box

The sub functions identified in 3.2 "Break-up into sub functions" and the mentally designed solutions for the sub functions are now clearly arranged in the morphological box.

	Sub functions	Solutions for the sub functions			
		1	2	3	4
A	Create rotation	Electric motor	Diesel engine		

⋮

... <Text> ...

This introduction guides the reader. So he cannot lose the overview across the sequence, which improves the understandability of the Technical Report.

The chapters "Introduction" and "Summary" are of major importance for the Technical Report. These two chapters are the first ones most readers scan after a quick look into the title and table of contents, before they start to read the text thoroughly. These chapters are introduced in the following overview together with examples of structure and contents.

Characteristics of Introduction and Summary

The introduction

– is located at the beginning of the text and is normally the first chapter.
– describes the starting situation at the beginning of the project, the relevance of the project for the particular field of science, the results of the research for the society, and other similar aspects.
– can contain a description of the project task and project target formulated by the author.
– can also contain thoughts related with the project regarding the following topics: economics, technology, laws, environment, organization, social care, politics, or similar topics.
– contains a short description of the technical surroundings and technical prerequisites for the execution of the project.

- should tie up with prior knowledge and experiences of the readers.
- strongly influences the motivation of the reader for thinking and inner dispute about the contents of your Technical Report.

The summary

- is located at the end of the text and is normally the last chapter.
- can have document part headings like: summary, summary and conclusions, summary and evaluation etc.
- discusses the task. Hence: What should be done and what has actually been reached, where have been special difficulties and which parts of the task could eventually not be treated and why.
- normally describes shortly what is covered in which chapter and subchapter of the Technical Report (picking up the structure!). When writing these sentences, you should express how the document parts are logically related with each other (starting with ..., then ..., next ..., due to ...).
- can give advice for a reasonable continuation of the project or scientific work in the conclusion. Such advice is normally based upon experiences made during the work on the current project.

3.1.4   List of References

The (overall) list of references is normally put directly after the last text chapter and lists all references to the literature from which there are citations in the whole Technical Report. In larger documents (e.g. in manuals or textbooks) there may be a capitular list of references after each chapter. The appendices follow the overall list of references or the capitular list of references of the last text chapter. It is unusual to integrate the list of references into the appendix. The list of references of a Technical Report has its own chapter number and stands alone between text and appendix or appendices.

To present all information regarding working with literature citations together, we have concentrated the why and how to make literature citations and how to design the list of references in Sect. 3.5.

3.1.5   Other Required or Useful Parts

The position of other required or useful parts within the Technical Report and their layout is even more dependent on university or company internal regulations than this is the case e.g. for the list of references.

A bound thesis like a diploma, bachelor or master thesis often contains the **task**. It is written by the institute or the supervisor and bound with the rest of the thesis. It is usually the first sheet after the inside front cover.

In addition, a diploma, bachelor or master thesis requires that the student presents a **declaration in lieu of an oath**. This declaration confirms that the student has written the thesis himself and that all used literature sources, rigs, machines and tools are listed completely and truthfully. The exact wording and the position of the declaration in lieu of an oath within the thesis are generally defined by the university. Beside to diploma, bachelor and master theses, doctorate theses and other final theses also contain such a declaration in lieu of an oath. The declaration in lieu of an oath must be personally signed by the candidate for the bachelor, master, diploma or doctorate degree. In most cases it is even required that the signature may not be copied. In bachelor, master and diploma theses the declaration in lieu of an oath is mostly part of the front matter and follows directly after the task, in doctorate theses it is mostly part of the back matter and is located directly before the Curriculum Vitae (CV).

In bachelor, master, and diploma theses there is often a page with **acknowledgements**. This is mainly the case, if the project, which the thesis describes, has taken part outside of the university and if there shall be expressed a special thank you to staff members of industrial companies. However, you should not forget the supervisors from the university here. Without their willingness to supervise the project and the writing of the thesis, and without their experiences, which topic is suited in which detail as a bachelor, master, diploma, or doctorate project, many projects would not take part at all or would last much longer than planned. If nothing else is defined or required, the acknowledgements should be put before the table of contents or the preface/foreword, if there is one.

Books often have a **foreword** or **preface**, pointing out the information target, changes since the last edition or specific rules how to use the book. A foreword or preface is always located directly before the table of contents.

Eventually, an **abstract** is required by the institute or company. In any case it may not be longer than one page, better is only half a page. Its heading is "Abstract" (or "Short summary") to distinguish it from the normal "Summary" at the end of the Technical Report. The abstract is put directly in front of the introduction. For most articles in scientific journals an abstract is obligatory before the article text following the title and the names of the authors. This abstract of an article is often in italic type and occurs in English and German or another language.

Therefore, the **front matter** consists of: title leaves, Task (What should be done?), declaration in lieu of an oath (How was it done? Alone!), acknowledgements, preface or foreword, and summary. Then the text chapters and list of references follow.

Often the Technical Report has **annexes**. Here the different indexes beside list of contents, list of references, glossary and index can be incorporated.

These annexes usually follow the list of references. Examples:

- List of figures
- Figures
- List of tables
- Tables
- Test and measuring protocols
- List of important standards
- List of abbreviations
- List of used formulas and units
- Bill of materials
- Technical drawings
- Manufacturer documents and other sources

Accompanying material like the bill of materials, technical drawings, measuring protocols, slides, models, leaflets, brochures etc. can be presented as independent appendix chapters or as subchapters in an appendix. They can be bound in the Technical Report or delivered in an external container like a second volume, a drawing roll, a folder or a box.

To place figures, tables and measuring protocols into the appendix/appendices is only useful, if placing them in the body would disturb smooth reading of the text chapters too much.

The appendix/appendices in the list above can be independent chapters. Then they get consecutively numbered appendix chapter numbers. If the number of appendix chapters becomes too large, the annex materials, lists, and accompanying materials can be combined in one common chapter "Appendix", see Appendix A in this book. The **Glossary**—if in opposition to ISO 7144 it is put at the end of the Technical Report—**and Index remain separate chapters in any case**, and they appear behind the appendix/appendices. The page numbers for the Glossary and Index may be just Arabic numbers as in this book or a combination of a letter for the chapter and a number (e.g. page numbers C-1 to C-10) or a combination of the chapter name and a number (e.g. page numbers Index-1 to Index-10).

The following two examples (left and right) demonstrate that an appendix can be organized as one chapter with subchapters or as several chapters. The variant "Appendix as one chapter" has the advantage that the document part numbers remain smaller and clearer. Besides the text chapters are normally subdivided, so that the chapter heading is the superordinated concept. It is more logic in itself to apply this approach in the appendix as well. Therefore, the structure on the right is recommended.

Alternatives for the structure of the appendix

Appendix as several chapters		Appendix as one chapter	
1 Introduction 3		1 Introduction 3	
(2-7 = other chapters)		(2-7 = other chapters)	
8 Summary and Conclusions 66		8 Summary and Conclusions 65	
9 References 68		9 References 67	
A List of Abbreviations A-1		A Appendix A-1	
B List of Figures B-1		A.1 List of Abbreviations A-3	
C List of Tables C-1		A.2 List of Figures A-6	
D Important Standards D-1		A.3 List of Tables A-47	
E Bibliography E-1		A.4 Important Standards A-65	
Glossary Glossary-1		A.5 Bibliography A-67	
Index Index-1		Glossary Glossary-1	
		Index Index-1	

Please compare the examples above with the structure of a Technical Report according to ISO 7144:

– **Front matter**: Front cover (cover pages 1 and 2), title leaf, eventually errata page(s), task (What should be done?), eventually declaration in lieu of an oath (How has it been done? "on my own"!), eventually acknowledgements, abstract (What is the result?), foreword or preface, table of contents, list of figures, list of tables, list of abbreviations and symbols, glossary.
– **Body**: text chapters with required figures, tables, formulas, list of references.
– **Appendix/Appendices**: annex material like figures, tables, bibliography, list of standards, eventually accompanying material, eventually glossary, index(es).
– **End matter**: curriculum vitae of the author, inside and outside back cover (cover pages 3 and 4), eventually accompanying material.

Now we want to give you hints for the design of the different parts of the appendix.

A List of figures contains figure numbers, figure titles, and page numbers. A List of tables contains table numbers, table titles, and page numbers. The page numbers should be placed right justified along a common building line. The distance between the end of the figure or table title and page number should be filled with leading dots as in the table of contents.

Figures or tables as annex material are often arranged in the appendix due to comfort reasons, because figures and tables can be handled more easily in an appendix than in a text chapter. Such an appendix forces the readers to turn the pages back and forth in the Technical Report very much. The text-figure-relationship of this solution is quite bad. Therefore, figures and tables should rather be integrated into the text chapters in the front.

However, design drawings should be placed in an appendix. If they would all be integrated in the text chapters, this would disturb smooth reading too much. If individual drawings are explained in the text in detail (e.g. Modification of an experiment rig), these drawings may additionally occur in the appertaining text chapter, eventually as a reduced DIN A4 or A3 copy.

The List of abbreviations contains the used abbreviations—in alphabetical order—and an explanation for each of them. The explanations should start at a common building line on the right side of the abbreviations. If the explanations have the character of definitions, they can be introduced by equal signs. If the abbreviations are a combination of the first letters of the words in the explanation, these first letters can be emphasized with bold type and eventually capital letters and underlining.

If a Technical Report contains many mathematical formulas, a List of used formulas and units may be helpful. The explanations of the formula symbols should again start at a common building line on the right side of the formula symbols. Depending on the type of your report you may as well create other lists, e.g. for checklists, exercises, link lists etc.

If you cite other materials (brochures by associations or companies, catalogues by manufacturers, important standards and other literature) as references, they must appear in the list of references. If they are not cited, it is sufficient to just add them to your appendix as additional information (as original or copy). Please plan this early and do not write notes into your original materials. You may mark selected dimensions with yellow highlighter. Highlighter in other colors create shadows when Your Technical Report is copied.

The pages of the appendices may get consecutive Arabic numbers (continued from the last page of the list of references). However, according to ISO 7144, the appendices get consecutive capital letters instead of chapter numbers and the page numbers consist of chapter letter and a consecutive Arabic number.

Enclosed documents or copies usually have their own page numbers. These existing page numbers remain untouched and every brochure, catalogue, or technical document gets a consecutive document number ("1", "2", "3", … or "Document 1", "Document 2", "Document 3" …), which is glued onto the document with white adhesive label. Sticky-notes ("Post-it") are not suited for this purpose.

Usually the enclosed documents are original brochures or photocopies, which are bound with the other pages of the Technical Report. If the enclosed documents cannot be bound, because they are too thick, too many or larger than DIN A4, they should be enclosed in a separate folder or box or roll. Such documents or drawings which are delivered separate from the Technical Report are listed in the table of contents of the Technical Report at the logically right position and get a comment like "(in drawing roll)" or "(in separate folder)". Example:

Reference to Appendices in drawing rolls and separate folders

...

8   Summary and Conclusions 66
9   References 68
    A   List of Abbreviations 76
    B   List of Figures 78
    C   List of Tables 81
    D   Important Standards 83
    E   Bill of Materials 84
    F   Drawings (in drawing roll)
    G   Manufacturer Catalogues (in separate folder)
    H   Bibliography (in separate folder)

If an appendix has a substructure, a title leaf for the appendix chapter is very useful. Such a title leaf gives an overview of the appendix and forms a capitular table of contents. For example, Appendix A contains a bill of materials and (bound) design drawings. In the front in the overall table of contents this structure would be displayed as follows:

Structure of the appendix in the table of contents
A   Drawings A-1
    A.1   Bill of materials A-2
    A.2   Assembly drawing A-3
    A.3   Component drawings: Gripping claws A-4
    A.4   Component drawings: Chassis A-9
B   Manufacturer catalogues B-1

In the back in the Appendix A the reader would be reminded of the structure of the appendix by means of a title leaf. This creates clarity and overview for the reader, Fig. 3.7.

▶   The layout of the title leaf of an appendix is partly equivalent to the layout of
    tables of contents (bold type of the chapter heading, leading dots, page
    numbers) and partly equivalent to the layout of title leafs (generous spread
    and pleasing arrangement of the printing ink on the paper).

If the appendix/appendices contain many plots, measuring protocols, program listings, drawings and other documents printed on DIN A4 paper, these **annex** materials can be bound separately as volume 2. The table of contents of both volumes should list all contents in both volumes, i.e. volume 1 and volume 2 have an overall table of contents with the sections "Contents – Volume 1" and "Contents – Volume 2".

Fig. 3.7 Example of a chapter
ToC for an Appendix

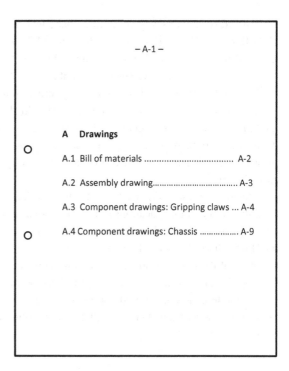

– A-1 –

A   Drawings

A.1 Bill of materials A-2

A.2 Assembly drawing.................................. A-3

A.3 Component drawings: Gripping claws ... A-4

A.4 Component drawings: Chassis A-9

If there are many files that belong to your Technical Report or project, you can insert a list with the path and file names and the file contents into the appendix. This list can be designed like a commented link list or a definition list.

Now some remarks regarding special appendices that—if they exist—must always be separate appendices following the other appendices or the list of references. They are listed in consecutive order.

A **glossary** contains technical terms and explanations of these terms. It is helpful, if the Technical Report deals with a specific field and the readers may not completely know the relevant terminology of this field. If the Technical Report is written in English but published in a country with an official language other than English, you should think of adding the technical terms in the official language of the country after the English term. You can add it in brackets or with a dash. To combine the technical terms in the official language with the technical terms in English facilitates the exchange of ideas in the scientific community, because much of the literature is written in English and many conferences are held in English. The technical terms in the glossary are accentuated by bold or italic type. They are ordered along a common building line. The explanations start either on the right side of the terms at another common building line or in the next line indented by approximately one or two centimeters.

An **index** contains keywords in alphabetical order. It is only useful for larger Technical Reports. The entries must be formulated *from the readers' point of view*. A structuring into superordinated and subordinated concepts (on max. two levels) results in more clarity and better overview. The page numbers are partly added with commas directly after the keywords or they are displayed right justified along a common building-line. The gap between index entries and page numbers should then be filled with leading dots. This looks more pleasing, especially, if the index has two or more columns. If an index entry occurs on more than one page in the text and on one page there is more or very important information, this page number can be accentuated by bold type and thus it can be marked as main entry.

Doctorate theses (Ph. D. theses) normally have a **Curriculum Vitae** (CV). It lists roughly the previous educational and professional development of the doctorate candidate or author. In doctorate theses another part in the appendices is the **Declaration in Lieu of an Oath**. Often it is placed behind the List of References or behind the CV. Contents and structure of the CV depend very much on the university or faculty. Also the placement of CV and Declaration in Lieu of an Oath might be different than proposed here. Therefore the doctorate candidate should ask his doctoral supervisor for an example CV and Declaration in Lieu of an Oath, eventually with anonymous (i.e. unrecognizable) personal data.

3.2 Collecting and Ordering the Material

Up to now the following parts of the Technical Report have been created:

– exact title with design of the title leaf and
– structure (detailed up to 10-point-structure or finer).

These parts give orientation for collecting the material. They help to answer the following questions:

– What is needed overall?
– What is already available?
– What is still needed?

Collecting the required material and information must be oriented towards the information target and the target group and answer the following questions:

– Which knowledge and experiences do I want to impart how?
– Which literature sources do I have to cite?
– Is there a visualization (figure, table, formula) for each important statement or at least a bullet list?

In most cases, to collect, order, and write something the reader does not need is unnecessary work!

Now we want to describe how the material collection can be done in practice. All spontaneous thoughts and ideas regarding the contents of the Technical Report should be collected on one or more sheets of paper regardless of their sequence and their assignment to already existing items in the structure. Eventually, you want to note each idea separately on note sheets, writing paper or file cards e.g. in DIN A5 landscape format. Then you should write on the front sides only, so that you can spread the sheets or cards on the desk or floor to sort and order them.

The material collection should include information, in which section of the own Technical Report which sources (internet pages and documents, books, articles and other references) shall be referred to or cited. Right from the beginning, please write down all required bibliographical data completely and the citations in the text exactly with author, year, page number(s), section number or exact URL (internet address and date when the information was accessed), so that when you start writing you do not need unnecessarily long time for (re-)searching what you had "digged out" before and you can cite correctly.

If a Technical Report is more comprehensive, i.e. it has more than 20 pages, the material collection should not be done for the whole report. In this case it is better to execute a separate material collection for each chapter or subchapter.

▶ Ordering the material can be executed in different ways. In this phase you should again have your target group in mind:

   – Are all noted items interesting for the readers? (if not: cross them out)
   – Can the noted items be already assigned to existing items of the structure? (If not: Open a new item in the structure)
   – If you took your notes one after the other on Din A4 paper, you should mark your ideas according to their "sequence". You can, for example, use the document part numbers from the structure and consecutive numbers within one section. If you noted each of your ideas on separate sheets or cards, you can order the sheets or cards according to their "sequence".

The "sequence" of thoughts can follow a chronological or logical order (by starting conditions, targets, alternatives, components, fields of knowledge, branches of science etc.). This chronological or logical order is—at least partly—predefined by the already existing structure of the Technical Report.

Therefore, ordering the material is only possible, if a 4-point- and 10-point-structure have been previously created. Due to the order of work steps recommended here, collecting and ordering the material will automatically be logical und oriented towards the information target. This method saves working time, because all work results fit together and only a few of them will end up in the "wastebasket".

3.3   Creating *Good* Tables

Tables display information in a matrix of rows and columns. The fields in this matrix are called cells.

In smaller tables the impression overweighs, that information can be displayed very systematically, well-arranged and structured in tables. However, larger tables are often confusing due to their poor amount of visualization. If the table displays words, this is less problematic than if it displays figures.

For readers, who read a figures-table, it is often difficult to estimate relations of numbers and to compare sizes. Therefore it is often required to visualize figures-tables by means of diagrams (charts).

Yet, you still have to proof calculations, statistic analyses, and experiment results with exact figures in your Technical Report.

▶   Therefore I propose the following procedure: Bulky and confusing
        figures-tables are provided in an appendix. In the text chapters there are well
        arranged visualizations (diagrams, charts) of the figures and the figure titles
        also refer to the related table in the appendix, e.g. with the annotation "(see
        table xx, page yy)".

In integrated office program packages the word processing and spreadsheet programs are compatible, figures (or tables resp.) can be easily exchanged between them. In such programs the figures in the table can be displayed in a selectable diagram type as presentation chart.

Usual and often-used diagram types are line, column, and circle diagram. Please refer to Sect. 3.4.5 to find tips for the selection of the diagram type and the design of schemes and diagrams.

3.3.1   Table Numbering and Table Headings

In the Technical Report tables have a table heading or caption which consists of table number and table title. In the presentation most people only use the table title (i.e. the text part of the table heading). In the table heading you may specify, which data is contained in the table, which basic conditions were relevant, which statement or conclusion the table shall proof etc. Moreover, cited tables must get an indication of source (citation) here.

The components of a table heading are called as follows in this book:

Table 16 Results of the fuel consumption measurements	table heading
16	table number
Figure 16	table label
Results of the fuel consumption measurements	table title

If the table is continued, the table heading may get a comment, that the table will be continued on the next page. This is useful, because the reader then knows of the continuation right from the beginning of reading the table. The following examples show this and other possible methods of informing the reader about the continuation. In the following examples the complete table is visualized by two short lines.

At first an example of a table which expands across one page only.

Table heading for table on one page

Table 16 Results of the fuel consumption measurements

Here is an example how a table can be labeled which expands across more than one page.

Table heading with continuation hint (method 1)

Table 19 Results of the welding experiments <to be continued>

Table 19 Results of the welding experiments <continued>

The table number and title is exactly repeated on all subsequent follow-up pages. The comments "<to be continued>" and "<continued>" show, whether the current table section is on a first page or a follow-up page.

If a more explicit labeling is desired, the comment "<to be continued>" can also appear at the end of the table:

Table heading with continuation hint (method 2)

Table 19 Res...	**Table 19** Res... <cont.>	**Table 19** Res... <cont.>
<to be continued>	<to be continued>	

Another way to label the table headings, if the table expands across more than one page, is to apply the numbering method of technical drawings for this purpose. Here is an example.

Table heading with continuation hint (method 3)

Table 19 Res... <page 1 of 2>	**Table** 19 Res... <page 2 of 2>

The table numbers can be consecutively counted through the whole report (example: 1, 2, 3, ..., 67, 68, 69). They can also be a combination of the chapter number and the table number which is consecutively counted within the chapter (example for chapter 3:

3-1, 3-2, 3-3, …, 3-12, 3-13). Instead of a hyphen you can also use a dot to structure the table number (example: 3.1, 3.2, 3.3, …, 3.12, 3.13). Books that have many small tables on one page sometimes use a combination of page number and table number within the page (example for page 324: 324.1, 324.2, 324.3).

The consecutive numbering of tables without chapter or page number has the advantage that the total number of tables can be determined quite easily. However, if you want to add or leave out a table later during the writing process, there are disadvantages, if you do not use the automatic numbering functions of your word processor, because all subsequent table numbers and the cross-references to the tables must be changed.

If you want to create the list of tables automatically, you should format all table headings with the same formatting template (style, paragraph format) and use this style only for table headings. Otherwise the automatic creation might be incomplete or might have undesired entries.

In general, the table heading shall describe the contents of the table as accurate as possible. Table heading, header, and introductory column shall be understandable without further explanations. The legend and eventually table footnotes may complete this information.

For the formulation of table headings the same rules are applicable as for the formulation of document part headings, see Sect. 2.4.3.

In the text there should be at least one cross-reference to each table which is integrated in the current text chapter. If possible, the table should be arranged near the cross-reference. If there are several cross-references from the text to a table, the table should be arranged near the most important cross-reference. Cross-references from the text to a table can be formulated as follows:

- …, Table xx.
- …, see Table xx.
- … shows the following table.
- … shows Table xx.

The table heading shall be positioned near the appertaining table. Therefore the distance above the table heading and below the table should be larger than the distance between the table heading and the table.

If your table heading expands across more than one line, all text lines of the table title start at a common building line on the right side of the table label.

If you copy or scan a table from a source of literature or you integrate a table from the internet as a graphic file into your Technical Report, you should cut off the table number and table title or you should not scan it together with the table. Then you should type your own table number and table title. The table title can either be taken over from the literature source or the internet without change or you formulate your own, new table title. All table numbers and table titles are created in the same way with your word processor. The result is a consistent overall impression.

Cited tables which you did not develop on your own must get a note of reference (citation) in the table heading. How this note of reference looks like is described in Sect. 3.5.4 in detail.

▶ You should plan early which tables you wish to incorporate into your Technical Report. Even if you will insert or cite a table later on, you should insert a table heading as a placeholder at the correct place while you are writing your text You can add a note such as „Attention:" or „Table comes later:" or „###" at the beginning of the table heading.

3.3.2 The Morphological Box—A *Special* Table

The morphological box is—beside its use as a creativity method—a central element in design methodology. There the morphological box is one work step in the design process. Hence it occurs quite often in Technical Reports.

Characteristic attributes of design methodology are thinking in functions and the sequence of work steps.

Starting from the list of requirements (requirements specification), the main function for the piece of equipment that is to be planned must be defined. It has proven to be practical to set up the list of requirements in the form of a table. Next the main function is divided into sub functions. The sub functions are always formulated according to the principle "execution on the object". Examples: create force, transform torque, guarantee steerability etc. Then design solutions are developed for each sub function.

In the following step the sub functions and the found solutions for the sub functions are clearly arranged in a matrix-shape in the morphological box, i.e. in a special table. The solutions for the sub functions can be displayed in the table cells only verbally (very often!), only graphically (with principle drawings) or verbally and graphically. From here on two different options how to proceed are possible, i.e. with several concept variants or one concept variant.

Different concept variants are usually marked in the morphological box with coloured or otherwise distinguishable lines. You can also use numbers or abbreviations. Please refer to Table 3.1 to see such a morphological box with several concept variants.

Then the concept variants are evaluated with respect to their technical and economical properties, Tables 3.5 and 3.6. The results of this technical and economical evaluation will be summarized in the s-diagram, Fig. 3.8. In the s-diagram you can recognize the best-suited concept variant.

In the evaluation procedure with **one concept variant only** a morphological box is arranged without marking concept variants. Then all design solutions of the sub functions are verbally evaluated. This means, that for all design solutions of the sub functions their advantages and disadvantages are listed in bullet lists. The design solution that shall be used is announced in the text in a separate line as follows "selected: <complete name of the solution of the sub function>". Then follows a short explanatory statement of about one to three sentences. The concept variant is now mentally combined and mounted from the best-suited solutions of each sub function. The marking of this concept variant in the morphological box follows at the end of the verbal evaluation, e.g. by shading the selected solutions of the sub functions in bold, see Table 3.2.

If a verbal evaluation of a sub function solution has only advantages and disadvantages are not specified (or vice versa), then you should express that with the word "none" or with the symbol "–". Now an example for such a verbal evaluation is shown for "Sub function C: lift water".

Verbal evaluation of a sub function solution	
Solution C2: Rotary pump	
Advantages:	– smooth
	– large delivery volume
Disadvantages:	none (or "–")

In the following you will get to know both variants of the morphological box (with several or only one concept variant) in a common example. Both variants of the morphological box refer to a fast turn-off device of a nuclear power plant, which has been developed at University for Applied Sciences Hannover for a design planning task.

In the first morphological box, Table 3.1, several concept variants are marked. In a Technical Report the different concept variants should be marked—if possible—with a colored line for each concept variant, because the readers can then distinguish the concept variants best. Colored pencil is better suited than other pens, because the color does not appear on the rear side of the paper, as this is often the case, if you use highlighter or felt pen. Sometimes this even happens with ball pens. Here, digits in brackets are used to distinguish the concept variants. In any case you have to undoubtedly assign the different colors or line types and line thicknesses or numbers or abbreviations to the different concept variants in a legend below the morphological box.

In the morphological box with several concept variants you mark the solution, which is best-suited after the technical and economical evaluation, with a much thicker line, if printing and duplication is in black-and-white. If you use digits or abbreviations, you may use boldface typing to mark the chosen concept variant.

In the morphological box with only one concept variant you mark the solutions of the sub functions, which are best-suited after the verbal evaluation, with a bold emphasis, Table 3.2.

Table 3.1 Morphological box (for a fast turn-off device in a nuclear power plant) with several concept variants, the electrical solution has been selected

	Sub functions	Solutions of the sub functions			
		1	2	3	4
A	Notice incident	**Electromagn. clutch (1) (3)**	Sensors (2)		
B	Disconnect device from normal operation	**Electromagn. clutch (1) (2) (3)**	Hydraulic clutch	Pneumatic clutch	
C	Lower control rods	**Electric drive (3)**	Hydraulic drive	Pneumatic drive (2)	Self- weight (1)
D	Connect brakes	**Electromagn. clutch (3)**	Hydraulic clutch	Pneumatic clutch (2)	Always connected (1)
E	Create braking force	Disc break (2)	Drum break (1)	**Induction break (3)**	Hydraulic damper
F	Transmit braking force	Cogwheel and cograil (1) (2)	Threadrod and nut	**Friction wheels (3)**	
G	Control braking operation	Centrifugal governor (1) (2)	Time controller	**Distance controller (3)**	
H	Enable end cushioning	Hydraulic oil brakes	**Hydraulic damper (1) (3)**	Pneumatic damper (2)	

Legend
Variant 1 = electrical-mechanical solution
Variant 2 = electrical-pneumatical solution
Variant 3 = electrical solution

Table 3.2 Morphological box (for a fast turn-off device in a nuclear power plant) with one concept variant

	Sub functions	Solutions of the sub functions			
		1	2	3	4
A	Notice incident	**Electromagn. clutch**	Sensors		
B	Disconnect device from normal operation	**Electromagn. clutch**	Hydraulic clutch	Pneumatic clutch	
C	Lower control rods	Electric drive	Hydraulic drive	Pneumatic drive	**Self-weight**
D	Connect brakes	Electromagn. clutch	Hydraulic clutch	Pneumatic clutch	**Always connected**
E	Create braking force	Disc break	**Drum break**	Induction break	Hydraulic damper
F	Transmit braking force	**Cogwheel and cograil**	Threadrod and nut	Friction wheels	
G	Control braking operation	**Centrifugal governor**	Time controller	Distance controller	
H	Enable end cushioning	Hydraulic oil brakes	**Hydraulic damper**	Pneumatic damper	

Special Design Rules for the Morphological Box

– The various sub functions in the morphological box usually have a different number of solutions of the sub functions. This results in empty cells at the right margin. These empty cells remain white, they are not shaded in gray or crossed out.
– Entries in the morphological box should always be left justified. If such entries are centered, this results in a too confusing layout.
– If you define several concept variants, you should give them meaningful and easy-to-remember names. Examples:
      articulated arm, rotatable, and portal roboter or
      hydraulic, pneumatic, and electric solution.
– In the following parts of the Technical Report these variants are always referred to with their once defined names and even sketches of the concept variants are labelled with these names as well. This is much better than names like Variant 1, Variant 2, Variant 3 etc., because you can remember names much better than digits and combine the names with inner images of the concept variants in your head.
– You should identify your designed solutions of the sub functions horizontally with numbers and the sub functions vertically with capital letters. This has the following advantages: If you address cell 3.2, it is not clear, whether this is the second cell in the third row or the third cell in the second row. However, if you apply the recommendation above, C2 is the second solution of sub function C. This unique name can e.g. be used in the verbal evaluation of the solutions of the sub functions.
–   If a sub function has several subgroups or characteristics, you can subdivide this sub function in the morphological box. The subgroups are then "numbered" with small letters. For example, sub function C "store water" (in a water purification plant) shall be subdivided in container number, type, shape, and size. In the morphological box, this has to be noted as follows: The solutions of the sub functions are addressed with a combination of capital letter, small letter, and number, e.g. Cb2 canister, see Table 3.3.
–   If the solution of a sub function needs to be subdivided into several subgroups or characteristics, this is not marked with another number or letter, but the term which describes the solution of the sub function expands across several columns and every subgroup gets its own Arabic number. In the next morphological box, two solutions of

Table 3.3 Morphological box: sub function with several subgroups or characteristics

	Sub functions	Solutions of the sub functions		
		1	2	3
C	Store water			
Ca	Number	Three	Five	Ten
Cb	Type	Barrel	Canister	Bottle
Cc	Shape	Cylinder with lid	Cuboid with handle	Cylinder with neck
Cd	Size	600 l	100 l	50 l

Table 3.4 Morphological box: subdivided sub function

	Sub function	Solutions of the sub functions				
		1	2	3	4	5
A	…					
B	Motor and cylinder cooling	Air cooling		Water cooling		
		Ring cooler	Tube cooler	Forced-circulation cooler	Flow cooler	
C	…					

sub functions are subdivided into subgroups: air cooling and water cooling of a motor. The subgroups each have their own number, so that they can be addressed without doubt with letter and number (B1 ring cooler to B4 flow cooler), see Table 3.4.

After the design of the morphological box is described, in the next section you will find information regarding evaluation tables.

3.3.3 Hints for Evaluation Tables

In evaluation tables several variants are evaluated based on different criteria. Examples are tables with criteria for the selection of a location of an industrial company, cost-benefit analyses, or the evaluation of concept variants according to VDI 2222 and VDI 2225 sheet 3. There the concept variants are first evaluated regarding their technical properties, then they are evaluated regarding their economical properties and then two evaluation tables are created.

At first, we want to show you the procedure as it is defined in the standards. Then we want to recommend a few deviations from the standard. Please speak with your supervisor or customer in advance, which procedure and table design shall be used.

In VDI 2222, the concept finding for technical products is standardized. It has the phases

– planning,
– concept-finding (list of requirements),
– function analysis (specification),
– concept (concept variants with technical and economical evaluation),
– draft (assembly drawing),
– optimization,
– refinement (single part drawings), and
– production of a prototype.

Beside many other examples the evaluation of a water purification plant is introduced.

The concept variants should get meaningful names, which your readers can keep in mind easily. The header should be emphasized with bold type as here in the book or with

Table 3.5 Technical evaluation properties of a water purification plant (acc. to VDI 2222)

Technical evaluation properties of the water purification plant	Points for variants 1 to 4				
	Var. 1	Var. 2	Var. 3	Var. 4	ideal
Blockage risk	2	3	4	3	4
Emission of smell	3	3	3	3	4
Emission of noise	3	3	2	3	4
Required space	1	2	3	2	4
Operational safety	3	3	4	2	4
Sum	12	14	16	13	20
Technical rating x	0.60	0.70	0.80	0.65	1

Table 3.6 Economical evaluation properties of a water purification plant (acc. to VDI 2222)

Economical evaluation properties of the water purification plant	Points for variants 1 to 4				
	Var. 1	Var. 2	Var. 3	Var. 4	ideal
Excavation	2	3	4	3	4
Concrete work	3	3	3	3	4
Expenses for pipes and fittings	3	3	2	3	4
Assembly costs	1	2	3	2	4
Maintenance costs	3	3	4	2	4
Sum	12	14	16	13	20
Economical rating y	0.60	0.70	0.80	0.65	1

gray shading. But now let us look at the individual steps of the evaluation, Tables 3.5, 3.6, and Fig. 3.8.

The strength s of the variants, which results from the x, y-coordinates of the four variants, is now drawn as points s1, s2, s3, and s4 into the so-called s-diagram. The ideal solution s_i is drawn at the position x = 1.0 and y = 1.0. Then a straight diagonal line from the lower left to the upper right corner is drawn that runs across the whole diagram. The best concept variant, here variant 3, is the one which can be found in the far right and far top of the diagram.

All evaluation tables must give exact information at first sight. If you read the table and the legend, it must be clear, which criteria have influenced the evaluation how strong and which variant could gain how many points. Basic rule: The evaluation table shall not be a mental exercise. Therefore, it must always be explicitly stated which variant has "won": the variant with the most points or—much more seldom—the variant with the least points.

For an appropriate evaluation you often have to give different evaluation criteria a different level of influence on the final rating of the concept variants. To express this different level of influence, weighting factors have been introduced. You have to multiply

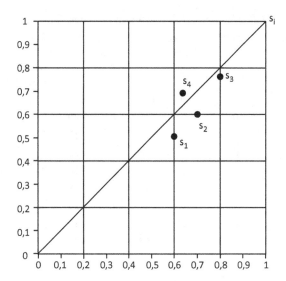

Fig. 3.8 Technical-economical evaluation of the concept variants of a water purification plant (from VDI 2222)

the simple point value with the appropriate weighting factor to get the total point value of the variant regarding the current evaluation criterion. Adding the total point values of all criteria leads to the sum total value of the current variant, which is listed in the evaluation table in the lowest row.

Table 3.7 is a concrete example, how an accurate evaluation table looks like. It has been created to evaluate several variants of the chassis of a boat trailer.

Please look at the legend. At first, all abbreviations are explained. Then the meaning of the evaluation factors is defined. So, the reader can comprehend, how each sum total has been computed. To avoid logical mistakes, you should apply the following principles when you give evaluation points:

- The given points are "positive points".
- High simple point values mean high value and high benefit for the users.
- Due to the multiplication with the simple point value high weighting factors result in a high level of influence of the current evaluation criterion on the rating result (=the sum total, the added total points) of the concept variant.

In the example "chassis of a boat trailer" the evaluation criteria self-weight and load carrying capacity must be weighted differently: If the load carrying capacity raises, the (positive) point value increases. But if the self-weight raises, then the (positive) point value must decrease! Written like a formula the situation looks as follows:

Table 3.7 Example of a technical evaluation table for the chassis of a boat trailer

Evaluation criterion	Weight	1 axis, 2 wheels		2 axes,4 wheels		2 axes, 3 wheels, 1 wheel steerable		2 axes, 4 wheels, 1 axis steerable	
	w	SP	TP	SP	TP	SP	TP	SP	TP
Self-weight	10	3	30	2	20	2	20	1	10
Assembly	6	3	18	2	12	1	6	1	6
Price	10	3	30	2	20	1	10	1	10
Ease-of-use	8	3	24	2	16	2	16	2	16
Design	4	3	12	2	8	1	4	1	4
Chassis	8	4	32	3	24	3	24	3	24
Load carrying capacity	8	2	16	4	32	3	24	4	32
Sum total			162		132		104		102

Legend
w = Weighting factor
SP = Simple points
TP = w · SP = Total points
w = 2, very low impact
w = 4, low impact
w = 6, medium impact
w = 8, high impact
w = 10, very high impact
SP = 0, not suited
SP = 1, with deficiencies
SP = 2, satisfactory
SP = 3, good
SP = 4, very good

Point values of evaluation criteria with high and low benefit

load carrying capacity ↑ ⇒ benefit for the users ↑ ⇒ simple point value ↑
self-weight   ↑ ⇒ benefit for the users ↓ ⇒ simple point value ↓

We can see that there are obviously two different cases.

Two weighting types in evaluation tables

1st case: parallel point values	measuring value of criterion ↑ and points ↑ (here: load carrying capacity)
2nd case: opposite point values	measuring value of criterion ↑ and points ↓ (here: self-weight)

Whether the points must be distributed parallel with the measuring value of the criterion or opposite to it causes many logical mistakes in evaluation tables. It is a problem of language logics, which is anyway harder to understand than mathematical logics for many people.

Often several of these evaluation tables follow one another. Then every table must get its own legend, so that every table is readable on its own and unnecessary turning of the pages is avoided.

A few keywords are a signal for opposite point values. Examples: High **effort** or **expenses** are the opposite of high benefit and have to be weighed with **opposite point values**. Raising cost expenditures result in decreasing simple points. A raising initial training effort, to learn how to handle a technical product is treated in this way, too. And finally a raising learning effort also results in a falling simple point value.

When distributing the simple points think of the users or customers and not of the manufacturer or service provider! Then it is easier to decide, whether parallel or opposite point values are required.

3.3.4   Tabular Re-arrangement of Text

REICHERT has described the method didactic-typographic visualization (DTV) in several books. In his approach continuous text is broken up, shortened, and visualized. This improves the clearness compared with continuous text. The result are either tabular arrangements of information or text graphics. Here is one example for a tabular re-arrangement of text:

Didactic-typographic visualization by tabular re-arrangement of text
Original text:

The sterilization temperature for sterilizing the tank should be at least 135 °C for 30 min. The sterilizing temperature at the condensomat should not fall below 125 °C.

Improved version:

Minimum temperatures for sterilizing:

location	temperature
at the tank	temp. = 135 °C, 30 min
at the condensomat	temp. = 125 °C

After you have got to know several rules and recommendations how to design figures-tables and text-tables, morphological boxes and evaluation tables and after a first short introduction to didactic-typographic visualization by tabular re-arrangement of text, in the next sections you will get hints for designing and creating figures.

3.4 Instructional Figures

Information can be displayed in different ways:

- with letters (words, sentences, text-tables),
- with figures (figures-tables and formulas) or
- as graphic display or figure resp. (diagram, illustration, graphic, image, scheme, etc.; a
  differentiation and definition of these terms will be provided later).

These different ways of presenting information have advantages and disadvantages. The verbal presentation of information is very exact and you can communicate abstract ideas, but it is not very clear. Figures are also very exact, but comprehensive figures-tables are also not very clear, even if you apply an optimal table design. If the figures are visualized in a diagram, the clarity is improved a lot.

Many other situations can be explained much simpler and clearer in figures than in verbal descriptions or tables and formulas. Figures are an eye-catcher, they are noticed first. Information, which is graphically displayed, can also be read and computed much easier.

Figures create associations in the reader's brain and thus activate the fantasy. Therefore, graphics are motivating and memorable. Partly they can even substitute detailed text descriptions. Figures emphasize structures and visualize the real life. Figures facilitate and intensify understanding processes. Therefore, information presented in figures can be remembered much better than read information. The efficiency of information transfer becomes better.

However, figures are not always the best way to present information. Abstract information can be better described with text.

▶ **Assignment of information to the text and to figures**

It is important to find a good mixture and assignment. The author decides,

- which information shall be displayed as text,
- which information shall be displayed only in a figure and
- which information shall be offered in both ways.

In this context, people use the term text-figure-relationship. This text-figure-relationship must be well planned. In addition, the placement of the figures relative to the text must be planned. By selecting appropriate positions for the figures and by giving cross-references from the text to the figures the author recommends a reading sequence.

The figures can either be integrated in the text or appear in a separate appendix. Both methods of figure placement have advantages and disadvantages. When revising the text the figures can be managed easier, if they are placed in an appendix, since they keep their position in the appendix when text changes are entered. If the figure would be placed in the text, this required additional page layout measures (move paragraphs from "below the figure" to "above the figure" and vice versa, adopting the cross-references from the text to the figure). Placing the figures in an appendix has one severe disadvantage. Because readers have to turn the pages back and forth, the text-figure-relationship is quite bad and the understandability sinks.

Therefore we recommend placing the figures near the appertaining text. Ideally, the figure is on the same page as the explaining text. If front and reverse pages are printed, the figure shall be on the same double-page as the text. If this is not possible due to layout problems, the figure can occur one page later, while a cross-reference from the text on the preceding page to the figure on the next page facilitates that the reader can turn the page at once and eventually several times.

▶   Always give your readers a timely cross-reference to figures and tables!

The easiest method to give such a cross-reference to a figure is to formulate any statement and add the cross-reference at the end with a comma. Example:

Cross-Reference to a figure

The force flow in the cup press runs from the plunger via the frame to the press table, Figure 12.

If you give such a cross-reference and the figure is not on the same side, the reader will turn the page and expect the figure there. With this very simple hint "…, Figure 12." the text-figure-relationship is assured and the understandability of the Technical Report is improved very much.

▶   You should draw or scan your figures as early as possible parallel with writing the text. If you start too late with creating the figures, you might be forced to work very fast (quick&dirty). Every reader can see this in the final product.

If you look at how the message is transported in figures, this leads to the overview in Table 3.8.

The further information about figures first deal with planning the figures. The basic rules for information-effective design of figures are explained and some rules for figure numbering and figure subheadings are introduced.

Other sections of this chapter deal with the displayed contents and the resulting type of graphic display. There you can find tips for creating (digital) photos, photocopies, scans, computer graphics and CAD drawings, schemes, diagrams, sketches, perspective drawings, technical drawings, mind maps and pictorial re-arrangements of texts.

Table 3.8 Systematic structure of graphic displays according to function and contents

Graphics		
Schemes (simplify reality)	Diagrams (explain abstract ideas)	Icons (create associations)
– symbol	– bar chart	– concrete pictogram (shows items or animals or persons)
– principle drawing	– point chart	
– block diagram	– line chart, e.g. graph of a function, nomogram	– abstract pictogram (shows abstract terms, must be learned)
– function scheme		
– flow chart of a process	– circular chart or pie chart	
– org chart	– area diagram	– symbol
– map	– body diagram	– logo
– wiring diagram, hydraulic diagram, pneumatic diagram, piping diagram	– Venn diagram	– traffic sign
	– tree diagram	– information sign
– technical drawing	– Gantt diagram	– etc.
– sketch for computations	– network plan	
– perspective drawing (incl. explosion drawing)	– flowchart	
– comic	– structogram	
	– mind map	

3.4.1   Understandable Design of Instructional Figures

As shown in Table 3.8 graphics or figures resp. have three possible functions. They shall either simplify reality (e.g. principle drawing, map), or explain abstract ideas by means of spatial arrangement (e.g. bar chart, pie chart, tree chart), or create associations (e.g. logo and pictogram).

Figures are often too little exact for these information purposes, but they have instructional advantages due to their similarity with the optically recognized world. In most cases it is especially effective, to present the overview information in figures and the detail information as text. To prevent misunderstandings and wrong interpretations of your readers, you should keep the following basic rules for figure design. If you keep these 13 basic rules, this is already a big step towards "good" figures.

---

13 Basic rules for information-effective design of figures

1. Accentuate important items!
2. Delete/leave out unimportant items! (use max. four to seven graphic elements in one picture, otherwise the picture becomes overloaded.)
3. Line thickness and font size must be sufficient! (The figure shall be readable without problems from the normal reading distance of 30–40 cm.)
4. The eye follows dominant lines. Therefore, relationships of graphic elements shall be emphasized (lines, arrows, columns, rows, common color). These relationships should also be specified in detail (What is the character of the

relationship? What does it mean?) by means of labels in the figure or explanations in the legend.

5. Spatial closeness of elements is understood as conceptual similarity (objects near to each other belong together).

6. Elements, which are placed above or below other elements, are interpreted as hierarchically super-ordinated or sub-ordinated. This emphasizes functional structures.

7. Elements, which are placed beside each other, are interpreted as time or logical sequence.

8. If the elements are arranged in a circle, this appears as a cycle, a frequently repeated sequence.

9. If one element surrounds another element, this is understood in such a way that the exterior term semantically includes the interior term.

10. Elements like boxes, bars, lines, columns must be clearly marked (either by text labels or by graphical explanations/pictograms).

11. One type of element may have only one function within one figure or figure series. For example arrows can be used for the following types of information:

   – direction of force,
   – moment of torque,
   – direction of movement,
   – flow of information,
   – cause and effect,
   – note etc.

   The different meaning of the arrows shall be visible due to a different graphical design of the arrows. So, double arrows are used to express a cause and effect relationship. If in the same figure you also want to draw a double arrow for a moment of torque, this arrow must look much different and the difference must be communicated to the reader.

12. Axes have large figures on the (vertical) y-axis at the top and on the (horizontal) x-axis on the right side.

13. For some diagram types, there are standardized symbols. Example: DIN 66001 for flow charts defines that a rectangle is used for an operation, a diamond for a decision and a rounded rectangle for start and end of a procedure. Other standards: DIN 32520, DIN 66261. Naturally speaking, such standards must be applied.

The application of basic rules 1 and 2 is also called "didactic reduction".

By leaving out and simplifying information, the reader can concentrate on what is important, following the motto "Something that is not shown cannot be misunderstood."

When designing figures, you can apply the following measures in addition to accentuate parts of the figure and to influence the reading sequence:

- **Color**: Colors can be remembered much better than section lining, different line thicknesses or line types. Red is the most popular and most striking color. Please use color very careful to avoid overshooting the target. And keep in mind: Colored objects and lines appear much different, when printed in black-and-white, because the printer changes colors to grayscale values.
- **Arrows**: There are many graphic design variants available for arrows. Arrows can point out important details. Arrows can also have other functions, see rule 11.
- **Oversize**: Details, which are not eye-catching, but important, are displayed larger and unimportant details are left out or displayed pale or blurry. The scaling factor of a detail within a normally (100%) sized environment can be raised up to 1.5 (150%) without irritating the reader.
- **Line thickness**: Important elements are drawn with 0.75–1.50 mm thick lines. Less important details appear with a line thickness of 0.25–0.50 mm.
- **Framing**: To accentuate important details, they can be framed or marked with a circle.
- **Detail enlargement**: A rectangular or circular frame is repeated in the same figure or (not so good solution) in a second figure (magnifying glass). The connection between the segment in original size and the enlarged segment must be recognizable (connecting lines). The frames must be similar (e.g. rectangle with constant width-to-height-ratio or circle) so that the reader can assign the segment of the figure in original size and the enlarged segment with each other.
- **Colored background or shading**: An important section of the figure gets a colored or gray background. In this case the contrast between object(s) in the figure and background must be sufficient.

▶   Please use only a few of these measures to accentuate parts of the figure and
       to influence the reading sequence. Otherwise, the figure might become
       confusing. Do not accentuate too many details. Otherwise, the influence on
       the reading sequence might get lost.

If you design your figures or modify figures from literature sources according to the 13 basic rules, also apply the measures to influence the reading sequence. Then your readers can recognize and interpret the message of your pictures much easier. Your figures become better understandable. Your message sent out as a figure reaches your readers much better and exactly in the intended way.

3.4.2 Figure Numbering and Figure Subheadings

According to ISO 7144 figure subheadings shall be placed below the figure. This is correct in the Technical Report, while in a presentation the figure title is placed above the figure so that you can see it on the projection pane without handicaps. Figure titles are indispensable, but often they are missing!

The components of a figure subheading are called as follows in this book:

Figure 16 Overview of production process variants	Figure subheading
16	Figure number
Figure 16	Figure label
Overview of production process variants	Figure title

The figure subheading in the Technical Report (or figure title in the presentation) tells in words what the figure shows and completes the labels of axes, sectors, bars, etc. It gives an indication of reference (citation), if the figure is cited. In the figure subheading you can specify for which conditions the figure is valid, which statement it shall support etc. The figure subheading should have the following layout:

Example of a figure subheading

Figure 16 Overview of production process variants

The figure label is accentuated by bold type printing, the figure title is printed with your standard font.

If a figure spreads across more than one page, you can apply one of the methods to mark the continuation which has been shown for tables in Sect. 3.3.1 analogously for the figure.

The figure numbers can be consecutively counted through the whole report (example: 1, 2, 3, ..., 67, 68, 69). They can also be a combination of the chapter number and the figure number which is consecutively counted within the chapter (example for chapter 3: 3-1, 3-2, 3-3, ..., 3-12, 3-13). Instead of a hyphen you can also use a dot to structure the figure number (example: 3.1, 3.2, 3.3, ..., 3.12, 3.13). Books that have many small figures on one page sometimes use a combination of page number and figure number within the page (example for page 324: 324.1, 324.2, 324.3).

The consecutive numbering of figures without chapter or page number has the advantage that the total number of figures can be determined quite easily.

The figure subheading must appear below the appertaining figure. Besides, it should be placed near the appertaining figure.

▶ Place the figure according to the following rules
(exact vertical distances may vary):

- The distance between text and figure is two blank lines.
- The distance between figure and figure subheading is (a half or) one blank line.
- Below the figure subheading, there are again two blank lines.

Due to the larger vertical distance above the figure and below the figure subheading, the reader can easily recognize that figure and figure subheading are a unit.

If a figure subheading spreads across more than one line, the second and all subsequent lines start at the building line of the figure title, Fig. 3.9. You can also layout figure subheadings, which belong to several small figures like this. The small figures, which belong together, are placed in a row and marked with small letter and closing bracket or only with a small letter.

Plan in time, which figures, shall be integrated in your Technical Report. Even if you want to draw or get a figure later, include a figure subheading rather early to avoid, that all subsequent figure labels and the appertaining manually created cross-references must be updated.

If you want to create a list of figures automatically, you should format all figure subheadings with the same (paragraph) style and use this style only for figure subheadings. Otherwise, the automatic creation might be incomplete or might have undesired entries. For details, please refer to Sect. 3.7.

Precise location references are important in figure titles. They help the reader of the Technical Report to imagine the real situation the figure refers to. A figure title with a too imprecise location reference is e.g. "Figure 6.5 Pressure measuring points P and temperature measuring points T". This figure subheading lacks an addition like "… in the cooling channel" or even better "… in the cooling channel of the die-casting mold".

For the formulation of figure subheadings the same rules are applicable as for the formulation of document part headings, see Sect. 2.4.3.

In the text there should be at least one cross-reference to each figure, which is integrated in the current text chapter. If possible, the figure should be arranged near the cross-reference. If there are several cross-references from the text to a figure, the figure should be arranged near the most important cross-reference. Cross-references to a figure can be formulated as follows:

− …, Figure xx.
− …, see Figure xx.
− … shows the following figure.
− … shows Figure xx.

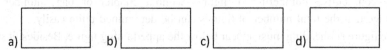

Figure 10 Overview of different methods of welding smoke suction which is not integrated into the welding process: a) pure hall suction, b) hall suction and hall ventilation, c) work table suction to the bottom, d) work table suction to the top

Fig. 3.9 Layout of figure subheadings that spread across more than one line and common figure subheading of several small figures

If you copy or scan a figure from a source of literature or the internet, you should cut off the figure number and figure subheading or you should not scan it together with the figure. Then you should type your own figure number and figure title. The figure title can either be taken over from the literature source or the internet without change or you formulate your own, new figure title. All figure numbers and figure titles are created in the same way with your word processor. The result is a consistent overall impression.

Figures which you scan, draw or copy from another publication (e.g. from a book, journal, newspaper, or CD) or from an internet page or a PDF file from the internet, must get a note of reference. How this note of reference looks like is described in Sect. 3.5.4 in detail.

It is quite often the case that you have to re-number the figures, because during editing the text of the Technical Report figures are moved to a different section, added or deleted. Then you have to find all figure subheadings in the complete text and all cross-references to figures. This can be executed with the word processor with the function "Find". You only have to enter the search string: "Figure_". The symbol "_" stands for a space character or a tab. Now you can check the figure labels and update them, if necessary.

There are several alternatives for the placement of figures: left justified, centered, or indented by a constant distance. If the figures are not too large and if this looks balanced, you should select the variant "figure title, left justified, figures start at the building line at the beginning of the figure title". This has the advantage that the layout is relatively smooth and the figure label is accentuated.

In the following sections, the various types of graphical illustration are introduced with their specialities. Then I will give you some hints how to insert photos and image files into your Technical Report and how to glue paper images into it.

3.4.3 Scheme and Diagram (Chart)

All images shall be structured as clear and simple as possible. The generally accepted rules of the current field of science and ISO, EN and DIN standards must be kept.

For several schemes (like flow chart, wiring diagram, hydraulic diagram, pneumatic diagram, piping diagram) appropriate symbols are standardized. Symbols for flowcharts are standardized in DIN 66001, for technical drawings some relevant standards are DIN ISO 128, DIN ISO 1101, DIN ISO 5456 etc. Other standards defining graphical symbols are: DIN 32520, DIN 66261.

For diagrams (bar chart, pie chart, curve chart etc.) there is the basic principle that all axes, bars, sectors etc. must be labeled unambiguously. Often the understandability of a diagram can be improved by clearly telling in the title above the diagram or in a label next to the diagram which statement the diagram shall support.

If a time-related process or development is displayed in a diagram, the horizontal axis is nearly always the time axis.

There are additional rules for the display of curve flows in coordinate systems. At first, the axes must be precisely labeled with

– physical value (as text or formula symbol) and measuring unit,
– measures (if it is a quantitative diagram) and
– arrows at or beside the axis ends (the arrows point to the top and to the right).

Figure 3.10 shows an example. The physical value can be a formula symbol (e.g. U in V) or text (e.g. Voltage in V). If you use formula symbols, the diagram can be used in a foreign language without change. If you use text, write it horizontally, because the graphic must not be turned. If necessary, the text can be hyphenated. If you cannot avoid a vertical text to specify the physical value, expand the letters of the vertical text (see DIN 461).

In diagrams with time progression the time must always be the horizontal axis. Proceeding time values begin on the left and go on to the right side.

The measures at linear scales (subdivision of coordinate axes) always include the value zero. Negative values get a minus sign, e.g. –3, –2, –1, 0, 1, 2, 3, Fig. 3.10. Logarithmic scales do not have the value zero.

If a diagram is used to derive measuring values from it, the application of ruled lines is useful, Fig. 3.11. According to DIN 461 the line width of the ruled lines shall have the ratio 1:2:4.

In diagrams with ruled lines two factors must be taken into consideration when it comes to defining how narrow the lines shall be: if the lines are too narrow, the reader is confused, but if the lines are too wide, it is very difficult to read off exact measured values. The scaling can be linear on both axes or logarithmic on one axis.

If a diagram is to be drawn with an interrupted scale or interrupted coordinate axis, there are two variants, Fig. 3.12.

If several curves shall be drawn in one diagram, the curves must have a clearly distinguishable labeling with short, clear terms and labeling letters and figures. The curves themselves must also be clearly distinguishable. To achieve that, you can use different colors, line styles and measured point symbols, Fig. 3.13.

The center point of the measured point symbols is placed in the diagram so that its coordinates equal the x-, and y-coordinates of the measured value. In diagrams with ruled lines, the curves and ruled lines can be interrupted before and after the measured point symbols, if this serves the clarity and exactness of reading off the measured values.

Fig. 3.10 Diagram with physical values, measuring units and measures

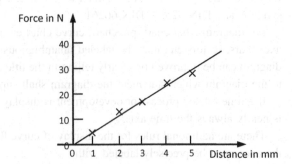

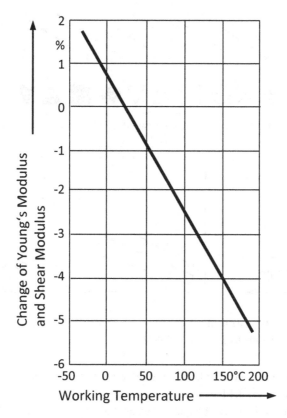

Fig. 3.11 Example of a diagram with ruled lines to read off exact values. *Source* Table appendix for the textbook ROLOFF/MATEK Machine Elements

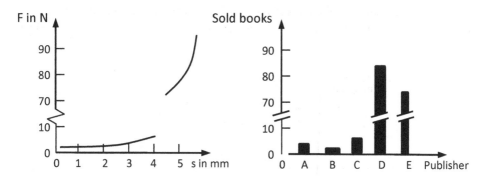

Fig. 3.12 Two variants to show that a scale or coordinate axis is interrupted

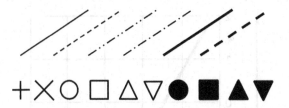

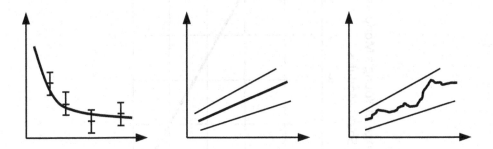

Fig. 3.13 Different, clearly distinguishable measured point symbols and line styles

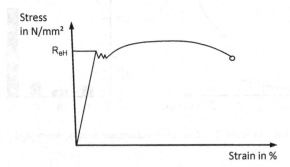

Fig. 3.14 Marking of the measured values and the limits of an error tolerance zone within the true value lies (the two left variants give the best contrast)

If you want to mark foreseeable or admitted error tolerance zones in a diagram, there are several variants, Fig. 3.14.

If a diagram shall only show the qualitative and not the quantitative relationship of two physical values, there are no scales on the coordinate axes. However, it is possible to mark important points by labeling their coordinates or physical value (again as text or as formula symbol), Fig. 3.15.

By selecting a different scaling density you can create a completely different optical impression using the same data and the same diagram type. In this way you can widely

Fig. 3.15 Example for a curve diagram that shows only the qualitative relationship of two physical values (stress-strain-diagram of a tensile test)

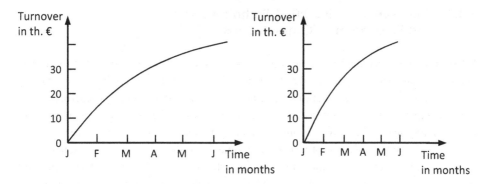

Fig. 3.16 Change of the optical impression of a curve by change of the scale density

manipulate the optical message of the diagram, Fig. 3.16. Please make sure, that you do not create an optical impression which represents the original data in a wrong way or which causes misunderstandings.

As already stated—to be sure to avoid misunderstandings the message or conclusion a diagram shall support or prove should be written above or beside the diagram or stated in the figure subheading.

As already stated in Sect. 3.3 presentation graphics (or diagrams or charts) are very well suited to display figures in a clear and easy-to-understand manner. The following table by MARKS (slightly modified) shows, which diagram type, can be used for which purpose, Table 3.9.

If the whole Technical Report shall be reproduced by copying, drawing and labeling the figures can be done with drop action pencil.

Table 3.9 Diagram types and their fields of application

Information	Diagram type			
	Pie chart	Column chart	Bar chart	Line chart
Development (of time)		(+)		+
Distribution (percentage)	+	+	(+)	
Comparison	+	+	+	
Frequency		+	+	+
Functional relationship				+
Comparison and development		+		+
Comparison and distribution	+	+		
Development and distribution	Separate charts			

Legend
+ Well-suited
(+) Less well-suited

3.4.4 The Sketch as Simplified Technical Drawing and Illustration of Computations

When designing technical appliances and machines you also have to deliver a computation of loads. This guarantees that the single parts can bear the applied loads without early failure. In these computations, sketches are used to display the geometrical situations, to explain formula symbols and to illustrate results of the computations. If possible, draw all physical values, which are computed in your equations. Here we show the following examples: diagram of forces and moments on the arm of a puller, scheme of a gearbox with exact specification of bearings, shafts and gears as well as simplified drawings of a motor, a screw and a cylinder. The sketch as illustration of a computation can also appear as perspective drawing.

Sketches in the Technical Report mostly have no figure number and no figure subheading. However, there are rules for sketches, too. They are derived from the basic rule:

▶   Sketch for computation of loads = intelligent reduction of technical drawing

During this intelligent reduction, you can leave out details, but never leave out important center lines. These are absolutely necessary for the quick recognition of part symmetries. Often it is also possible to imitate simplified displays from manufacturer documents.

When you draw machines or plants or parts of them for handling and manufacturing parts you should always draw the handled or manufactured part into your sketch. For example, it is hard to imagine that in a lifting device for the quantity production of a part exactly this part is missing in the sketch. For example, think of a cask claw. In the sketch for the computation of loads, the cask is drawn in red and eventually in a different line style. If the transported part is accentuated like this, the readers can understand the operational processes of the machine, the diagram of forces and moments, the computed loads etc. much better.

Sketches are also used e.g. in assembly instructions, to explain the spatial conditions. Here is a text example from an instruction for use of a lawn raking machine by TOP-CRAFT, which would be quite hard to understand without sketches (or photos):

> Insert bent fastening pipes into case holes. Put anti-vibration rubber disk between case and fastening pipes. Then fix draught relief of cable at fastening pipe. Now the lower push bow is stuck onto the fastening pipe.

Sketches for computations of load are always placed near the appertaining computations. They are always reduced to the essential information, Fig. 3.17.

It is very important to use unambiguous names and labels! If necessary, you should use short explaining texts beside the names and symbols of physical values and points to mark the position in the sketch unambiguously to which the computation refers, Fig. 3.18.

Fig. 3.17 Arm of a puller with diagram of forces and moments

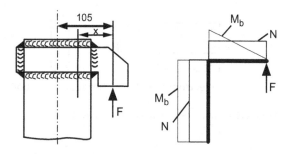

Fig. 3.18 Gearbox scheme with exact specification of bearings, shafts and gears

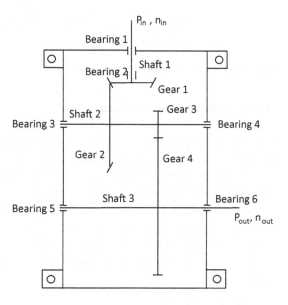

You may imitate or copy simplified technical drawings, e.g. from manufacturer documents and standards, if the clarity is not influenced, Fig. 3.19.

Sketches for computations of load are sometimes perspective drawings. Then the rules and tips in Sect. 3.4.5 can be applied accordingly. If the report shall be duplicated by copying later, it is also possible, to draw sketches for computations of load manually with drop-action pencil. Therefore, you can easily correct the sketches.

Eventually, figure titles below the sketches are useful to prevent unnecessary searching. As an example Fig. 3.20 shows various material profiles:

In general, you should apply measures like cross-references, figure subheadings, labels etc. to make sure, that the reader can follow your Technical Report without having to search much and without questions. Imagine your target group as engineers, who have no detailed knowledge of the current project and who shall still understand the report without feedback from you.

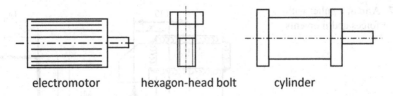

Fig. 3.19 Examples of simplified technical drawings from manufacturer documents

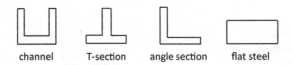

Fig. 3.20 Simplified display of the sections of different semi-manufactured bar materials

3.4.5 Perspective Drawing

Parts can be displayed in normal three-plane projection (orthogonal projection) or in a perspective projection. The perspective projection is easier understandable, Fig. 3.21.

In DIN ISO 5456 the following perspective views or perspective projections are distingished: isometric projection, dimetric projection, cavalier projection, cabinet projection, and central projection with one, two or three projection centers.

In mechanical engineering and electrical engineering the central perspective is seldom used. It can be found more frequently in architecture, civil engineering and design. The following list shows the perspective projections (all but central projection) in a comparison.

- **Isometric projection**
  Scale: B:H:T = 1:1:1, edges: $-30°$, $30°$ and $90°$. Linear dimensions can be directly measured parallel with the axes in all three axis directions.
- **Dimetric projection**
  Scale: B:H:T = 1:1:0.5, edges: $-7°$, $42°$ and $90°$. Linear dimensions can be directly measured parallel with the axes in two axis directions only.
- **Cavalier projection**
  Scale: B:H:T = 1:1:1, edges: $0°$, $45°$ and $90°$. Linear dimensions can be directly measured parallel with the axes in all three axis directions, radii and diameters in front view only.
- **Cabinet projection**
  Scale: B:H:T = 1:1:0.5, edges: $0°$, $45°$ and $90°$. Linear dimensions can be directly measured parallel with the axes in two directions only, proportions can be estimated well.

Fig. 3.21 Advantage of
perspective drawings: the shape
of objects can be determined
much easier

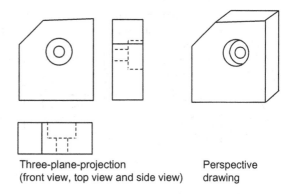

Three-plane-projection   Perspective
(front view, top view and side view)   drawing

In cavalier projection, objects seem to be very distorted to the human eye. Therefore, it is less suited for Technical Reports. The cabinet projection and the cavalier projection are the easiest to be drawn, since here the projection axes only have the angles 0°, 45° and 90°. These angles can easily be created with a set square (drawing triangle) and they can be drawn easily in freehand technique on normal 5 mm squared paper.

In paper shops they have special drawing board, drawing paper and graph papers like millimeter paper, logarithmic millimeter paper and isometric millimeter paper as well as different lettering and drawing stencils for manual drawing. Since in isometric projection circles are not projected to all views as circles, there are various ellipse stencils and special instruments which facilitate drawing ellipses.

In CAD programs you can create cutaway drawings, which—for example—present a look onto ¾ of the case of a drilling machine and into ¼ of the case to look at the gearbox inside. For creating exploded views, there are specialized (commercial) programs like Iso Draw.

Perspective drawings have the following advantages and disadvantages:

- clearer, easier-to-understand, better overview,
- saves space compared with three-plane-projection (3 views),
- supports the spatial imagination,
- but the drawing costs more effort than for three-plane-projection.

3.4.6 Technical Drawing and Bill of Materials (Parts List)

Nearly all design and projecting reports have technical drawings, often in a drawing roll or drawing folder. They are an important part of this type of Technical Reports. Therefore we will give you a list of frequent mistakes in technical drawings, see also Fig. 3.22.

Centre lines are often forgotten in technical drawings and in sketches. All parts which are rotationally or axonal symmetric must get center lines. That is also valid for holes, indexing circles, pitch circles, etc.

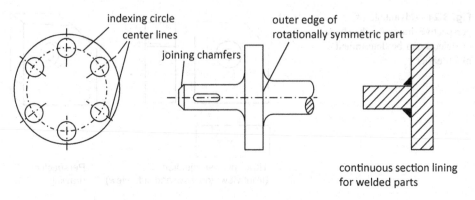

Fig. 3.22 Avoidance of frequent mistakes in technical drawings

If for example holes are regularly arranged at a circular flange, they get a common **indexing circle** and each hole gets a short center line, which cuts the indexing circle in normal direction. The center lines run perpendicularly to the indexing circle. It is wrong and not according to the standards, to mark each hole with a horizontal and a vertical center line.

Rotating edges of objects (bearings, bearing caps, etc.) are often forgotten, especially in section and assembly drawings. Therefore, after finishing your drawing, check all rotationally symmetric parts (with the bill of materials), whether all required edges of the objects are drawn.

Joining chamfers are also often forgotten. The remark "all edges which are not especially marked are broken" is not sufficient! Think about how the sub-assembly or assembly or device can be mounted! Without joining chamfers bearings, shaft seal rings, bearing caps etc. can only be mounted with a larger loose fit and then they usually can no longer fulfill their function. The check "fictive assembly of the parts" also helps to avoid the additional mistake, that a bearing cannot be mounted, because e.g. a shaft and a gear wheel are manufactured from the solid and the gear wheel is in the way.

Single part drawings must have **all dimensions** which are required to produce the part. It is quite frequent in assembly drawings, that some assembly dimensions are forgotten. Here are the most important assembly dimensions: maximum length, width and height (=minimum inner dimensions of the transport container or box), shaft heights, diameter and length of the end parts of the shafts for connecting other parts, index circle or hole distance(s) and hole diameter of flanges to fix the assembly, including flange thickness (because of the required length of the fastening screws, handle lengths, ball handle diameter (where the hand of the user is "connected" to operate the machine or device).

For assembly drawings of **welded parts** with section linings the following rule applies: welded sub-assemblies are continuously sectioned (not each plate with a different sectioning), because at the time of assembly the sub-assembly is one part.

The **bill of materials** (part list) is integrated into the Technical Report and added to transparent originals or plots of the drawings in a drawing roll, drawing folder, or drawing box.

3.4.7 Mind Map

Mind Maps are used for structuring topics, problems, plans, discussions etc. For example, you can draw a mind map during a presentation or lecture instead of writing the usual continuous text script. Mind maps also support brainstorming processes. In a mind map all aspects of a topic that must be considered become visible as main branches, which are branched to smaller twigs and in the end have detailed topics, questions, or aspects as leaves.

On the market there are many computer programs to create mind maps very fast and easy. For the following example, Fig. 3.23, the quite price worthy program Creative MindMap by Data Becker was used. In the mind map branches and twigs can be created quickly and a branch can be moved together with its twigs to a different location of the topic tree etc. More expensive and more professional software like MindManager have more comprehensive clipart libraries and more features.

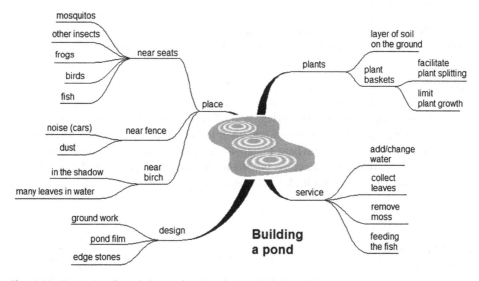

Fig. 3.23 Example of a mind map for planning the building of a pond in a garden

3.4.8 Pictorial Re-arrangement of Text

A text graphic consists of text, which is layouted with typographic measures like indentations, bullet lists, bold print etc. and graphical elements. These graphical elements are, for example lines, rectangles or circles, which are arranged "above" the text. The result is a smooth transition towards a diagram or chart. It is sufficient to use only a very limited number of these graphical elements to create a text graphic from a small amount of text. Here is an example as an inspiration for your own text graphics, Fig. 3.24. Also, Fig. 3.26 is a pictorial re-arrangement of text.

3.4.9 Creating Paper Images and Graphics Files and Incorporating Them into the Technical Report

So far we discussed the types of graphical illustrations, i.e. the "What". Now we will discuss the creation of the graphical illustrations and their integration into your Technical Report, i.e. the "How". But before you start, you have to make some preliminary decisions.

Preliminary decision 1: paper based or digital?
Printed photos are still used today, e.g. for metallographic micrographs in damage analyses, for photos of plants, as press photos etc. However, digital photos, scanned images and images from the internet are used much more frequently.

If you want to use colored images from books and brochures in your Technical Report, scan the image and print the page on a color printer. Later you can reproduce this page of your Technical Report with a color copier. Colored photos can be glued into each copy of the Technical Report. The handling of black-and-white graphics with brightness gradients and black-and-white photos is similar.

Paper based images can be manually created drawings of all types including sketches used as illustration in computations, mind maps, comics, design drafts etc. They are used whenever speed and creativity are important. However, in the meantime the usage of graphics and CAD programs is the standard.

Preliminary descision 2: use of graphics and CAD programs?
Using CAD programs saves a lot of time when you modify an existing part, when there are repeat parts and part families. When using CAD programs, the effectivity is much dependent on the task. CAD programs are often too complicated for casual users. They have so many functions, that you have to train and practice using the program. That is also true for graphics programs. Therefore, the advantages and disadvantages of graphics and CAD programs will be shown in the following overview.

Fig. 3.24 Examples of pictorial re-arrangements of text

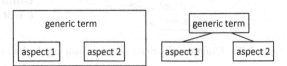

Advantages and disadvantages of graphics and CAD programs

Advantages:

- The graphics look tidy, there is a consistent overall impression.
- By using fill colors and fill patterns, area graphics can be created.
- Changes can be done quick and easy. (You should use a graphics program, which allows drawing on different levels—also called slides or layers.)
- The graphics can also be integrated in word processing or desktop publishing (DTP) programs. This enhances the Technical Report optically.
- If you use CAD programs, a perspective 3D display and the different views can be easily created once the object data has been entered.
- Employers expect that you have learned how to use standard software (graphics as well as CAD programs) during your study course or in an autodidactic approach. Therefore the time you invest in learning to use these programs is not lost.

Disadvantages:

- The time and effort you have to invest in learning complex graphics programs and for drawing PC graphics is quite high (depending on the program). The user interface is sometimes confusing and the time to get first results is so long, that drawing the figures with a simpler program is the better solution for the moment. Pixel graphics programs as well have different concepts for the user interface. If you have to work a lot with pixel graphics (e.g. with images from digital cameras or with GIF and JPG files for the internet/intranet), learn the creation and editing of these files early enough.
- Especially scanned figures sometimes can have different scaling factors in x and y direction. In addition, if you do not mention it, circles are no longer round! It is better to assign the scaling factors numerically.
- If after scaling your figure there are Moiré effects (checkered or striped pattern), undo the scaling and test scaling factors which are a multiple or a fraction of the original size by the factor 2 (factors 25, 50, 200% etc.).
- If after scanning a figure with a scanner there are Moiré effects, check that with the zoom function of your graphics program. Perhaps it is just a screen display problem. Moreover, it helps sometimes, to put a thick glass plate onto the scanner and to put the figure to be scanned on top.

No matter which software you use—sometimes it lasts quite long to create PC graphics, but you will be rewarded with good-looking results. The following overview shows important facts about graphics and CAD programs.

▶ You should decide early, which graphic you want to create with which pro-
    gram. Probably this decision will be influenced by possible license fees and by
    which programs are used in your working environment. Then you should learn
    how to work with these programs well before the end of your project, if that is
    necessary.

An exact estimation how long you need to create digital drawings is normally harder than it used to be when drawing by hand. Therefore, we recommend the following:

▶ Create graphics files as early as possible. Take into consideration that it lasts
    longer than you estimated, even if it was a generous estimation! In pixel
    graphics programs, using image effects can cause unexpected results. Test the
    functions of your graphics program on a copy of your data. Also, try out the
    integration of the graphics into your word processor or DTP program and the
    printed results as early as possible.

The next figure is a vector graphics file created with a graphics program, which looks attractive due to fill colors and fill patterns, Fig. 3.25. In the original work the author had drawn this illustration by hand with drawing ink and added the fill patterns with scratch foil.

▶ If the drawings shall look proper even in detail, you should zoom-in into tricky
    areas from time to time and check whether your graphic still looks tidy in the
    enlargement. Often you cannot see areas, which are drawn untidy in 100%
    scaling on the screen, but the printer will show everything much more precise.

Also create test printouts of your PC graphic from the word processor program. If possible avoid that in your word processor or in a graphics program several objects are lying on top of each other in one level, e.g. by using a different graphics program which enables working on different levels (layers). If several objects are lying on top of each other, you have to move the objects lying in the front to be able to select and edit the objects in the back. Then you need to reposition the objects in the front as they were arranged before. Grouped objects sometimes have to be ungrouped and grouped again later etc. All this is time consuming but inevitable without level management.

Preliminary descision 3: vector or pixel format? Which file type?
CAD programs internally use objects and create vector graphics files. Graphics programs either create pixel graphics files (the image consists of pixels) or vector graphics files (the image is a collection of objects).

If a line drawing is saved as a pixel graphics file, it is quite large, because for every single pixel the brightness and color must be saved. However, pixel graphics files are well

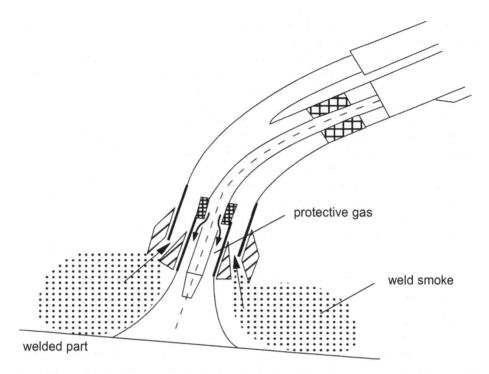

protective gas

weld smoke

welded part

Fig. 3.25 Section through a protective gas welding pistol with integrated weld smoke suction (fill patterns and arrows show the direction of flow of protective gas and weld smoke)

suited for halftone images (with continuous color and brightness gradients) and for scanned photos.

Vector graphics files use a different concept. If on a large area a circle is to be displayed, a pixel graphics file would become larger with increasing diameter. The vector graphics file does not change its size very much, because the only data stored in the vector graphics file are center point coordinates, radius and line thickness, style and color. For a straight line, the only data stored are starting and end point coordinates and line thickness, style and color. These graphics files are much smaller. Therefore, vector graphics files are better suited for line drawings than pixel graphics files.

File formats for pixel graphics files (also called raster graphics) are e.g. BMP, GIF, JPG, PNG, and TIFF. The three formats GIF, JPG, and PNG can be displayed by internet browsers and imported by most word processors. Almost all vector graphics file formats are vendor specific. Many vector graphics programs can read and/or write DXF. Standardization efforts derive from the design and manufacturing area (CAD/CAM). They use DXF, VDA-FS, IGES, SAT, IFC, and STEP as system neutral file formats. In the field of technical documentation, SVG is quite frequently used, because it can be displayed in internet browsers as well.

In the following important differences between the formats GIF, JPG, PNG, and TIF are described.

GIF files can only display a limited number of colors. They are well suited for pictograms and figures without color and brightness gradients. Another advantage is that you can combine several single GIF files to an animated GIF file which runs like a small film.

JPG files are well-suited for figures with color and brightness gradients. If you want to display large images on the screen, the format JPG offers to save the JPG file with the option interlaced. If you look at the image online, the figure is built up different. At first the 1st, 3rd, 5th line etc. of the image are displayed so that you can see a rough outline of the figure quickly. Then the 2nd, 4th, 6th line etc. are displayed so that the image is completely displayed. JPG files do not need much disk space, but it occurs quite frequently that JPG files are blurred, especially screenshots.

Since both formats GIF and JPG have disadvantages, the format PNG is also used. A detailed description of the file format is published on Wikipedia. PNG is an abbreviation of Portable Network Graphics. PNG has an image compression without losses, it supports different color depths and transparency. The image compression rate of the PNG format is normally higher than the compression rate of GIF files. However, PNG has the following disadvantages: more complex than GIF; no animation; not as high compression rate as JPG, but without losses; no support of the CMYK color model, which is required for four-colour printing, therefore no substitution of TIF files.

Avoiding distortion
No matter which format you use, please make sure that your images are not distorted. Distortion happens, if you use a different scaling factor in x and y direction. You can best avoid these problems, if you do not rescale your images with the mouse, but via numeric entry of a scaling factor. This factor should result from a multiplication or division by 2 (25, 50, 200% etc.). Please make sure, that the option proportional scaling or maintain proportions is switched on. (The different word processors, slide presentation and graphics programs call this a bit different, but the sense of the function is always the same.)

Selection of an adequate image resolution
If you scan images or create digital photos, you should plan early with which resolution you want to work. For a simple screen display GIF or JPG files with 72–75 dpi are sufficient, while for high-quality journals the resolution must be at least 300 dpi and TIFF files are preferred rather than JPG files.

▶   In the Technical Report 150 dpi resolution is a good compromise in most cases.
     And the higher the resolution, the larger is the resulting file size. This goes on as
     follows: If you insert many large image files without reference into your text
     files or slide presentations, your files will also become much larger.

The image resolution of scanned figures should not be too high. The image resolution for images, which shall only be displayed on a screen, is 72 or 75 dpi. In the Technical Report an image resolution of 150 dpi is sufficient for proper printouts in most cases. For four-color printing of journals and books, it must be min. 300 dpi.

Always save image files independent of and within text or presentation file
Image files, which you use in your Technical Report or in a slide presentation, should always be stored on the hard disk drive or a USB stick as separate graphics files, so that you can enter changes and apply all functions of your graphics program later on.

▶   It is best to save the graphics files in the vendor specific standard format of your graphics program and in the format, which you used to import the graphic into your text or presentation file.

If you stick to this procedure, you can also save additional information like copyright hints, image source, thematic context, entry into the list of references etc. for each graphics file. If you use the figures in a different context, like a different Technical Report or a different presentation, you can reuse this information.

Tips for Taking Photographs
In the following overview, some well-tried rules for creating photos in Technical Reports are introduced.

Rules for the design of impressive photographs
Picture detail selection

- "Close" to the object or people (leave out distracting information, emphasize details).
- Select unusual perspectives (worm's-eye view, bird's-eye view).
- If the size of the displayed item is not imaginable for all readers without problems, other items should be shown for a size comparison, which are well known to the readers. Examples: Ruler, man, hand, fingertip, banknote, coin etc. If you just specify a measure (e.g. 1:25) most readers cannot correctly estimate the dimensions.
- Try out portrait instead of landscape format and vice versa.
- Emphasis of the front part of the image provides depth (a plastic impression). This is especially nice, if the sides and the top part of the image are emphasized (view through an "archway").
- Accentuations with colors make the image more vivid. Use complementary colors red + green, blue + orange, yellow + violet, but: depending on the target group not too colorful.
- Shaking hands, handover of documents, cheques, certificates and similar situations should be photographed from the head to the mid of the thigh (above the knee).
- Take enough time (wait for better light conditions, search other standpoints).

Light and shadow

- Light from the side enhances the contrast.
- Reduce backlight by covering the sun.
- Backlight plus automatic exposure without backlight correction results in black objects or people in front of a colored background (only silhouettes).
- That can be a desired result.
- Backlight and a flashlight for the front part of the image result in a harmonic light distribution.
- In interior rooms, covering light sources with bright cloth or transparent paper and reflecting light with white areas can improve the light distribution.
- The flashlight shall not be directly reflected back from the object or person to the camera. Such reflections distract the viewer very much form the intended photo motif and sometimes outshine important details of the image.
- Digital photos are often far too dark. In interior rooms, you should use artificial light.

Creating photoprints yourself

If you develop black-and-white-photographs on your own, you can use rasterizing film for the photo process. These rasterizing films are exposed together with the negative (a little longer exposure time). The light creates an already rasterized enlargement of the negative on the photo paper, which can be copied in black and white with an excellent quality.

Creating paper images by painting, drawing, cutting them out or by copying

From childhood, we learn to paint and draw. Here it is important to use the right pencils, see Sect. 3.8.3. The manually painted or drawn images are normally glued onto a carrier page of your Technical Report and copied. Partially it comes still across, that images are copied from books or journals and glued onto a carrier page of your Technical Report, e.g. if they cannot be borrowed from a library. Also, cutting out paper images from leaflets or journals is sometimes necessary. Try to get two copies in this case—one for your personal archive and one for the paper original of your Technical Report.

When images are glued onto carrier pages and copied, some problems may occur, that need further treatment, e.g. undesired edges of the copied images. To modify photocopies you should use a magnifying glass and a drop action pencil with a soft lead, so that you can erase faults easily. Often i-dots, arrowheads and very thin lines (especially section lining) is too small or too bright in the copy. Emphasize them by redrawing! Your modified copy will be the new original and your readers will be grateful, if you invest time and effort here.

If a copied or scanned image has reference lines and labels naming important figure elements, there are often terminology problems. In the text of your Technical Report you are using one consistent term for the same item. In the scanned or copied image, another term is used. Either the source is written too much in general-language or it is too theoretical and uses too many special-language terms, which you do not want to use. The situation is similar, if you want to cite from a publication, which is written in a foreign language.

Removing an undesired term from the figure is one possible option. However, if the term shall be completely eliminated from the figure, removing the appertaining reference line is not always easy (e.g. when the reference line crosses many closely neighbored lines or cross-section linings). In this case, it is better to cover the undesired term with the desired term (manually by gluing or with a graphics program). It is not so easy-to-read, but acceptable, to explain the undesired term with the desired term in a legend.

▶   It must be avoided in any case that a figure contains special terms, which have
    not been introduced so far, and that there are different names for the same
    items or procedures in the text and the figure.

Didactic reduction on paper images, photos and pixel graphics
Sometimes figures, that you want to copy or scan, are a little overloaded. Cover not so important details (with white tape or labels or photocopy fluid) and accentuate important details (e.g. with arrows or circles).

In digital photos and other pixel graphics, you can cover unimportant details with white rectangles or lines and weaken not so important parts by making them more bright and blurry with your graphics program. This is called "didactic reduction".

Reserving Space for paper images
If a figure shall be glued in, there must be enough blank lines in vertical direction so that the figure fits in well. The space left white for the figure can be easily measured with the vertical position information of your word processor (line, column) or with the vertical ruler.

If you prefer paper based working, create a page with your word processor with your usual paragraph format settings. You should write consecutive numbers beginning with 1, one number per line. If you print that file, the numbers 1, 2, 3 etc. appear one below the other on the „line ruler". Now you can take a ruler, measure the height of your figure, read from the line ruler how many empty lines that is, add two lines and enter this number of empty lines into your text file.

You can find more details about gluing paper images onto carrier pages and labeling them in Sect. 3.8.3.

Inserting graphics files
When integrating graphics and scanned images into your text files, you sometimes have to change the figure size. You should always enlarge or reduce the size proportionally. In the graphics program, the relevant option is called "Maintain proportions" or "Width-to-height-ratio" or "Aspect ratio". After importing the graphic into your word processor or presentation program, do not pick the graphic with the mouse at a corner and draw it to the desired size, but click the graphic and use the function Format—Object, tab

Size or a similar way, where you specify the scaling as a numerical value. Otherwise distorted text labels and ovals instead of circles and other undesired effects can occur.

Copying figures with color and brightness gradients
If you have to create a larger number of copies, you can reproduce the figures with color and brightness gradients (halftone images) on a black-and-white copier. You should definitely use the photo key to avoid quality problems. If there are still quality problems even with the photo key, you have to rasterize your figures.

To achieve that you can use rasterizing film. The film is laid on top of the original during the copying process. The film should only cover the figure, so that the text does not lose its contours and contrast. The rasterizing film disseminates the halftone image into individual pixels. If prepared like this the copy of the page can be used as original for the further reproduction. Rasterizing film for photocopiers is quite expensive, but the quality of the rasterized halftone images is very good. They can be copied without problems.

▶ Try out all required work steps (test printouts and black-and-white copies of your printouts). If the image quality is too low, you can modify the image creation procedures, and you still have enough time to keep your deadline.

3.5 Literature Citations

When writing Technical Reports there are various versions of literature citations and lists of references. Therefore, it will be described in detail how to cite literature and to write a list of references. However, here are at first a few introductory remarks.

3.5.1 Introductory Remarks on Literature Citations

These introductory remarks shall begin with a definition of the term "literature citation". A citation is using a statement written by another author either literally or in one's own words and precisely documenting the literature source from where the statement is cited.

That means, citing is the same as copying text from someone else. In the context of "returned" doctorate theses they often use the term plagiarism. The process fo copying is legitimized and refined only, if you indicate where the copied information comes from.

The short note in the text, where the literature comes from is called citation. In the back in the list of references the cited publications are listed with their bibliographical data.

In scientific works, there are nearly always literature citations. Also in Technical Reports, which are partly not so strictly following the rules, there are often citations of statements written by other persons.

Literature citations have the following tasks:

- They help to describe the current state-of-the-art.
- They support the author's opinion.
- They emphasize the scientific character of the Technical Report.
- They exculpate the author from the responsibility for the contents of the citation (but not from the responsibility for the selection of the citation).
- They point out the author's accuracy and intellectual honesty.
- They underline the author's authority and credibility an.
- They permit that the readers can check the facts (statements, values in computations etc.) and by studying the literature from several authors to get a higher level of expertise in the current field of science.

Presenting the bibliographical data of cited literature helps the readers to borrow the literature from a library, to find it in the internet, or to buy it from a book shop. Therefore, there must be sufficient information for each publication.

If every author would use his or her own systematic to present the citations and the required bibliographical data, there would be a chaos. Therefore, there is a standard, that defines how literature citations must look like: ISO 690 "Information and documentation - Guidelines for bibliographic references and citations to information resources".

3.5.2   Reasons for Literature Citations

The skilled combination and collection of information from various sources of literature combined with correct citing is a central point of academic work.

Correct citations are a proof that the author knows the rules of scientific working. The selection and quality of the literature citations document how intensively the author has read about the "state-of-the-art" and the established theories. Thus, literature citations do not express, that the author has not had own ideas. On the contrary, it is absolutely necessary to make correct literature citations to facilitate that interested readers can get access to a field which might be new to them.

Correct literature citations emphasize the honesty of the author. It is part of a positive human and scientifically correct behavior, that literature citations are marked. If you copy texts or figures from other authors into your Technical Report without naming the literature sources, you pretend that the work results of other authors are your own. That offends national and international copyright laws and good manners within the scientific community.

Someone who knows the field well, in which you have worked, will quickly recognize the spots where you have used thoughts of other authors, but without correct literature citation.

In the meantime, there are even computer programs that help your supervisor or customer to check the text of your Technical Report. The programs delivers a more or less precise analysis, which parts of the text are identical with other texts published in the internet and thus helps to recognize plagiarism much faster. However, these programs are not perfect. It is recommended to use several of these programs and to evaluate the analysis by an expert.

No one wants to come into such a situation that he or she is accused of plagiarism. Therefore, the next section is a short description, which bibliographical data you have to collect for citations according to ISO 690. Then there are sections describing how literature citations must be made in the text and in which form the bibliographical data must be listed in the references. A further section deals with literature written in foreign languages.

3.5.3 Bibliographical Data According to ISO 690

ISO 690 contains rules for collecting lists of references. They describe which bibliographical data must be listed for the cited publications in which order. The layout of the list of references is in block format—a very compact form, which is used in many books and journals. In the standard, there is no information how the list of references is presented in the classic three-column form, which is usual in Technical Reports.

The term "publication" in a narrow sense contains textbooks, contributions to host documents, articles in journals or newspapers, company literature, brochures, catalogues, etc. In a wider sense, the term "publication" also contains the following media: record, radio transmission, video or TV film, computer program, documents that are saved on CD-ROM, personal messages as well as any information that is available via internet or intranet. In this book, we want to call all these publication types "literature" and "source".

The next two sections show how the citation in the text has to be made and how the list of references must look like.

3.5.4 Citations in the Text

How a citation is marked in the text, depends on what is cited and whether the citation is literal or analogous. We have to distinguish:

- literal citation of text
- analogous citation of text
- exact copy of a figure or a table (scanned, photocopied, redrawn)
- analogous citation of a figure or a table (adopted to the own information target by means of modifications).

Every publication from which information is cited, must get a citation in the front in the continuous text and an entry with all bibliographical data in the back in the list of references. It is not allowed to copy five articles from journals, put them into a plastic pocket, and label the pocket with "Brochure 5", if the five articles thematically belong closely together, but have been published in different journals and written by different authors. It is also not allowed to type a citation directly behind a document part heading and to hope, that the readers will understand that!

All literature should be at first listed in the back in the list of references. In the next step the cited information in the front should be typed into the text.

For large projects with intensive literature work the following work technique has proven to be practical: For all cited publications you should not only note the bibliographical data, but also unique identifiers (ISBN for books, ISSN for journals, SICI for articles in periodicals, DOI for electronic publications, LCCN for publications registered by the American Library of Congress, ordering number of the publisher etc.), the library where you found the publication, location within the library, and signature.

Now you can give back the literature and eventually borrow it again later, if you need to improve your text or forgot something in the list of references. Therefore, you do not need so much space for the books, and you do not need to extend the loan period after four weeks. If you borrow the books again, you have all required data available without having to search the books again. That saves much work and eventually reminder charges.

Every literature citation consists of the cited information and the citation in the front and the bibliographical data in the back in the list of references, Fig. 3.26.

When you cite text there are some rules, which are listed at first and then explained with examples. We introduce variants. Please select one variant and apply it consistently throughout your Technical Report!

Referencing scheme
The short note near the cited information in the text shall enable the readers to identify the corresponding entry in the list of references or in a footnote or endnote without doubt. Possible variants are listed below, examples are marked in bold and italic print:

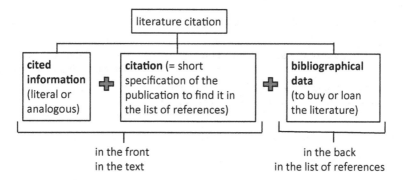

Fig. 3.26 Components of a literature citation

Basic referencing schemes:

number or symbol written in superscript, referring to a footnote or endnote (=note referencing)	[23]
short citation: own literature number in rounded or square brackets (=parenthetical referencing)	[18] or (18)
long citation: usually with author/s, year and page/s (=author-date-system, Harvard style of referencing)	[MILLER, 1993, p. 27–30]

The marking of citations with angular brackets "[]" is often used in textbooks and similar publications and we recommend it here. If you use round brackets "()", like proposed by ISO 690 and often used in the Anglo-American language area, the short citations (only using the literature numbers) could be mixed up with equation numbers.

First part of long citation (author part):

author name	SALINGER
author name and first name(s)	WINTER, John
author name and abbreviated first name(s)	HENSON, D. STIEG, MF. HALDANE, JBS.
two or three authors	MILLER, P. and JOHNSON, D.
four and more authors	MILLER, P. et al.
not mentioned in the standard: several authors with identical first names can be distinguished with Roman numbers	KLARE, Donovan I and KLARE, Donovan II
If a publication has no named author, a short name of the publishing institution with the location of the head office in brackets or one or more words from the title are used as author name. The letters N. N. are not used according to ISO 690.	BYU (Provo) DIERCKE atlas of the world
If not only the author, but also the publisher is unknown, a phrase can indicate that.	Publisher unknown Chinese proverb
optional: specification of the author's function	MILLER, Peter (ed.) or WARNCKE, Tilo (interviewed)
optional: author name(s) in CAPITAL LETTERS or small caps is more distinctive compared with the continuous text	

Second part of long citation (date part):

year of publication	2003
several cited publications of one author in the same year are distinguished with small letters	2003a, 2003b, 2003c
estimated year of publication	ca. 1920 ca. 1920 (copyright) ca. 1920 (printing)

Third part of long citation (location part):

optional: page number(s) of the source	pp. 27–30
optional: specification of a document part	subchapter 9.3

The citation is enclosed in round brackets according to ISO 690. In Technical Reports, it is often enclosed in square brackets to distinguish literature from formula numbers. Square brackets are recommended by several institutions relevant in natural sciences and engineering like ACS, AIP, AMS, CSE, ASME, and IEEE. Here are some examples of complete citations in the parenthetical referencing scheme:

Examples of long and short citation

[18]
    [KUHN, 1986a] etc.
    [18, p. 50]
    [HAMSING, 1993, p. 27–30]
    [HAMSING, 1993, section 2.7.3]

In the examples above, two methods of citing literature have become clear:

(a) only specification of the literature source (in the note referencing scheme and also in the parenthetical referencing scheme, when you specify only the literature number)

(b) specification of the literature source with the author, year of publishing and with or without the exact location in the source (page numbers/document part)

Method (a) is often used in Technical Reports and according to ISO 690 it is correct. Method (b) is nearly always obligatory in the humanities. Method (b) is called author-date system or Harvard referencing style. If there are exact page numbers or document part numbers, all parties concerned, i.e. the author himself, his supervisor, and the other readers can find the cited information faster.

Sorting in the List of references

There are various ways to sort the literature sources in the list of references:

– **Numerically by their order of appearance in the text**
The first cited source gets the number [1], the second source the number [2] etc. Then the list of references is not sorted alphabetically. This way of numbering is suited, if in the document there are only a few sources cited, e.g. in short Technical Reports or in journal articles. It is e.g. recommended by IEEE, ASME, and ICMJE.
– **Numerically by the alphabetical order of author names**
In larger works, the list of references in the back is in most cases sorted by the alphabetical order of author names. The literature sources do not appear in the text in the numerical order [1], [2], [3] etc., but e.g. in the sequence [34], [19], [83] etc.
– **Alphanumerically by the alphabetical order of author names, literature identifier contains first letter(s) of author name and number**
This method is not listed in ISO 690, but the ordering in the form [M1], [M2], [M3] etc. or [HAM93] saves time and effort in larger Technical Reports and it is recommended by AMS. If a source of literature is added or left out late in the writing process, only those literature numbers with the same letter(s) must be changed, not all subsequent literature numbers. In the end the alphanumerical identification can be replaced by a numerical identification.
– Please contact your supervisor or customer and arrange which method is to be used!

Alphabetical order of author names
In large alphabetically sorted lists of references there is sometimes the question, which is the alphabetically correct order of the author names. The following rules apply:

– Several authors with the same surname are ordered by their first names.
– Several publications of the same author are sorted by the year of publication (2008, 2006, 2005). If the citations do not only contain the literature numbers, but author and year or author, year, pages, identical years get small letters to distinguish the publications (2006a, 2006b, 2006c, …).
– Several publications of the same author with several co-authors are sorted by co-author names.
– If the co-author names are identical, but their first names differ, co-authors are sorted by their first names.

These sorting rules are applied accordingly, if there are even more identical sorting attributes, until a useful sorting attribute is found. Not mentioned in ISO 690: If there is no useful sorting attribute, use Roman numbers as a distinction.
Author names with "Mac…" are treated as author names with "Mc…".

Secondary citation
Citations from a primary literature source (citation from original work) must be specified as listed above, e.g. in the system "author, year, pages". If a literature source is hard to get or not available at all, but cited from another author, whose work you have, you can make a secondary citation. The citation must look like this:

> **Structure of the long citation for a secondary citation**
>
> [author (not available), year, pages cited by: author (available), year, pages]

Both publications are listed in the list of references.

Naturally speaking, if the secondary literature source only cites author and year of the primary literature source, you can only copy these two pieces of information.

> **Secondary citation**
>
> In the front in the text there is e.g. the following citation: [KLARE, 1963, 1974/75; TEIGELER, 1968 cited by BALLSTAEDT et al., 1981, S. 212].
>
> In the back in the list of references you have to list the three primary sources by KLARE and TEIGELER and the secondary source by BALLSTAEDT.

Document part headings without citation

After document part headings there are never citations.

Longer citations (several cited paragraphs)

If a single literature citation is long, that means it has several paragraphs and—beside the quotation marks—cannot be distinguished from the normal running text, the citation must be repeated at the end of each paragraph of the cited text. If the citation stood only once at the end of the last paragraph of the cited text, this would be an uncertain situation. The wrong conclusion would be possible, that the specification of the literature source only refers to the last paragraph. If you mark the literature citation by italic printing or an indentation, you can write the citation only once at the end of the cited text.

Literal citation

Each literally citied fact is marked with quotation marks.

> **Complete literal citation**
>
> "Therefore written queries (...) have been sent to all professional associations of technical writers in the world and to INTECOM."
>
> [HERING, 1993, p. 338]

Left-out information

If you want to cite only a part of a sentence or you do not cite a few words in the middle of a sentence, this must be marked with an ellipsis.

> **Literal citation with an omission according to ISO 690**
>
> "Therefore written queries (...) have been sent to all professional associations of technical writers in the world and to INTECOM."
>
> [HERING, 1993, p. 338]

According to DIN 5008 the ellipsis consists of three dots without parentheses and no space characters between the three dots.

According to ISO 690 the ellipsis consists of three dots with round parentheses and no space characters between the three dots.

According to Wikipedia the ellipsis must be enclosed in angular brackets [...] in citations.

According to www.thepunctuationguide.com the ellipsis should consist of three periods and no brackets. Each period should have a single space on either side, except when adjacent to a quotation mark, question mark or exclamation point in which case there should be no space.

Typographic accentuations in the source
If in the source text there is a typographic accentuation, you have to copy that in a literal citation exactly. This holds true for bold print, italic print, use of capital letters, indentations etc. In the example, the title of an article is literally cited and the italic print is used exactly as in the original article. If the italic print would have been omitted, the result were a distorting mistake.

Typographic accentuations in the source are copied in the citation

DILLINGHAM writes in his article *"Technical* Writing vs. Technical *Writing"*, that the knowledge of the technology is necessary for technical writers, so that they understand the products, they shall describe. This understanding of the technical details is the prerequisite of successful user information. [DILLINGHAM, 1981]

Comment on the cited text
If you do not want to copy the typography of the source or give comments regarding the cited text, you have to inform your readers, that you have changed the typography or presented your personal opinion. The easiest way to do that, is to use the text "note from the author:" and enclose the note in angle brackets <...> or angular brackets [...].

Note from the author
..., also called the European way <note from the author: typographic accentuation in the source text was not copied>.

Quotation marks in the source
If in the original text there are quotation marks, this must appear as half quotation marks, ...'in the literally cited text.

Quotation marks in the source become half quotation marks in the citation

"In this context it is especially remarkable, that he <note from the author: Leonardo da Vinci is meant> has introduced a completely new kind of visual display already around 1500 to the technical documentation. This is the 'explosion drawing', ..., which can be found e.g. in nearly all current vehicle spare parts catalogues."
[HERING, 1993, p. 18].

Analogous citation
Information that is cited analogously is not marked with quotation marks. It is only marked with the citation at the end of the cited text.

> **Analogous citation**
>
> WIERIGER assumes, that the gas drive is more ecological [12].

Positions of a citation within a sentence

The citation can appear at different positions within the sentence. Preferably, it should be integrated so that it disturbs the reading flow as little as possible. However, if misunderstandings are possible, where the cited information comes from, the citations must be placed so that the relations become clear.

> **Different positions of a citation**
>
> The physical basics have already been examined by SIMON [17]. He found out, ...
> This conclusion could be affirmed by physical [12, 17] and chemical [9, 22] experiments.
> Similar research [2, 7–10, 15] shows, ...
> Taking into consideration the findings of SMITH [16] and RIEMERS [9] ....

If short citations are applied (only using the literature numbers) like "[23]", this citation should not stand alone on a new line. Try to shorten the text before the citation so that the citation jumps up one line or add a few words before the citation.

Cited numerical values

Data and information, you use in computations of load, error calculations etc., must be cited. Examples: physical constants, material constants, computation procedures by manufacturers, standardized measuring procedures.

Citation of figures and tables without changing them

If you copy a figure or a table from a source without changing the contents, there are three ways of presenting the citation. As seen for citing text, you can type the normal citation. That is, you specify the number of the literature source in the list of references or the author's name, year of publication and eventually page numbers or section. Again, we recommend square brackets. In any case, the citation appears behind the figure subheading or table heading.

Quite frequently the note "(Source: author, year, pages)" with rounded brackets is used. If you have enough space, it looks more balanced to write this note left justified with the rest of the figure subheading or table heading on a separate line, see the example "Figure 23" below. In case of figures from brochures or corporate publications or company homepages you can also use the note "Corporate photography:" plus company name. Examples:

Figure subheadings for images cited without changes		
Figure 13	Section through an Otto motor [15, p. 50]	
Figure 15	Different combustion chamber geometries [18]	
Figure 17	Fuel consumption be with different combustion chamber geometry and equal cylinder capacity [MILLER, 1990, S. 35]	
Figure 19	Octane rating depending on Benzol concentration [MEHRING, 1992a]	
Figure 23	Percentage of lightweight construction materials in self-weight of formula 1 cars from 1950 to 1995 (Source: LEHMAN, 1995, p. 81)	
Figure 28	Different designs of needle bearings (Corporate photography: FAG)	

Citation of figures and tables with changing them

If you copy a figure or a table from a source with changing the contents, you should use the note "acc. to <author>" or "according to <author>" to express that it is an analogous citation. Your changes could be the following:

– leaving out/changing labels, so that they fit to the terminology of the text,
– leaving out details from the figure, and
– leaving out not so important rows or columns from a table.

Figure subheadings for images cited with changes	
Figure 13	Viscosity of different engine oils depending on the temperature according to PFITZNER [12]

Alternatively, you can give your readers a note in the figure subheading or table heading like "…, simplified compared with the source" to make clear, that it is not an exact copy, but that you have changed, left out, improved, or further developed something.

Citations from the internet

If you use information or an image from the internet, you have to cite it correctly. We recommend that in the last line of the figure subtitle you cite the internet address of the image (a) or of the HTML page on which the image is used (b). The examples show a result of a google image search. The internet address has to be entered to the list of references as well. There it has to appear together with an author and the date when the citation was recorded/when the figure was seen.

If the internet address is too long, you may shorten it so that the reader comes near the relevant topic on the cited homepage. Then please specify, where the reader has to click to reach the cited page. Alternatively, you write a shortened address or just the start address of the homepage in the figure title, and the complete address of the figure in the list of references.

Also, if an HTML page uses frames (you can recognize that when the internet address stays the same after you clicked a link, but the contents of the page has changed), please specify, where the reader has to click, to reach the cited page.

Examples of literature sources for images copied from the internet		
Figure 25	Working principle of a parallel plate capacitor (Source: www.vtf.de/p90_1_3.gif, visited on Aug 12, 2007)	(variant a)
Figure 25	Working principle of a parallel plate capacitor (Source: www.vtf.de/p90_1.shtml, visited on Aug 12, 2007)	(variant b)

The same rules for specifying the internet address of the files you found apply for textual information and other file types like PDF, Word, Excel, Powerpoint, MP3 etc.

▶ Please consider that the presented contents in the internet can change very quickly. If you found a valuable source, you may wish to copy the texts, figures, and files, which you might use later, to your hard disk drive. Then you should note, where and when you found the information (i.e. URL/URI and date, when you saw the information).

For cited figures, audio files, PPT presentations, texts etc. you can note the bibliographical data for a literature citation in a TXT file having—beside the extension—the same file name as the file with the contents to be cited.

3.5.5 The List of References—Contents and Layout

In the list of references the bibliographical data of the cited publications is collected. The list of references is always located directly after the last text chapter, normally after „Summary and conclusions". The numbering of the chapter "References" is according to ISO 2145, see also Sects. 2.4.2, 2.4.4, and 3.1.2. In articles, there are sometimes just intermediate headings without document part numbers. These headings are printed in bold type and usually with a slightly larger font size. The heading for the list of references is also "Works cited" or "Literature" or "Bibliography".

It might be useful, especially for larger Technical Reports to put up a structured list of references. In the structured list of references different types of literature are separated by intermediate headings, e.g. (a) Books, articles etc., (b) Laws, standards, regulations etc. and (c) Internet links see reference part in this book. You may also sort the list of references by topic. Then there could occur titles like "Literature on building construction", "Literature on underground construction" etc.

At the end of a structured list of references there is sometimes a section "Further links to the internet", "Further literature", "Further reading" or "Bibliography". Here you can list sources of literature, which have not been cited, but which are important for the treated topic: standard text books, literature on special topics, homepages of companies etc. The section with the bibliography can get an own document part number or—better—just a heading which is emphasized by bold printing and a slightly larger font size.

In very large Technical Reports you can also find the variant of a space-saving capitular list of references. The capitular list of references is located at the end of each chapter. The corresponding subchapter headings might be "1.7 References in chapter 1", "2.9 References in chapter 2" etc. The layout of capitular lists of references is similar to the layout of the list of references in journals. An example can be found below under "space-saving form". In each chapter, the numbering of the sources of literature starts again with "[1]".

Before you enter your first source of literature to your list of references, you have to define, which layout you want to use. There are three options: the first one is the classic three-column form, the second is the two-column space-saving form, which is also used in journals or third the block format according to ISO 690.

Figure 3.27 shows how to display the bibliographical data in the list of references in three-column and two-column form. This layout is suited, if you use short citations.

The following example shows the display of the bibliographical data in the list of references in block format. This layout is suited, if you work with long citations.

Layout of the list of references in block format according to ISO 690

Koller 1985 KOLLER, Richard: *Konstruktionslehre für den Maschinenbau – Grundlagen des methodischen Konstruierens.* 2. Auflage. Berlin, Heidelberg: Springer, 1985
Riehle/Simmichen 1997 RIEHLE, Manfred; SIMMICHEN, Elke: *Grundlagen der Werkstofftechnik.* Stuttgart: Deutscher Verlag für Grundstoffindustrie, 1997

The block format according to ISO 690 is even more compact. The following example shows the layout.

classic **three-column form** (in Technical Reports):

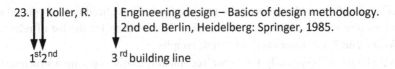

space-saving **two-column form** (in journals and books):

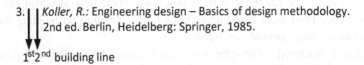

Fig. 3.27 Layout of the list of references in three-column and two-column form

Block format according to ISO 690
Graham, Sheila. College of one. New York: Viking, 1967.
– The real F. Scott Fitzgerald thirty-five years later. New York: Grosset & Dunlap, 1976.

The bibliographical data shall allow the readers of your Technical Report, that they can find the literature, which you cited in libraries, order it from publishing companies, industrial companies, and institutional bodies or buy it in a bookshop. It is a matter of fairness, that all required data is provided correctly and completely. You do not need to explicitly specify addresses of publishing companies, industrial companies, and institutional bodies, because their addresses can be found in publically available sources like Yellow Pages and the internet. You can find the address of publishing companies also in the impress of a journal by that publishing company or in the internet.

The list of references has three blocks of information, which correspond to the columns in the list of references in three-column format. In the first column, there is the running number of the source of literature. This column should be right justified. The numbering of the list of references can be equal as in the text, e.g. [1], [2], [3] or /1/, /2/, /3/or (1), (2), (3) etc. This is conform to DIN 1422, part 2, but this is not recommended in ISO 690. ISO 690 recommends the numbering scheme "1., 2., 3. etc."

In the second column, the author names are listed. There are the surnames of the authors with usually abbreviated first names or the long citation (Table 3.10). Academic titles of the authors are left out. The publications may have been printed by institutions instead of authors. Then the entries in the second column change, see Table 3.11.

The third column of the list of references is reserved for the bibliographical data. The sequence and structure of these bibliographical data depends on the type of literature and the referencing style.

A number of organizations and institutions have created styles to fit their needs. Publishers also often have their own in-house variations. Table 3.10 lists some popular referencing styles in various fields of science.

If a book has not been written by one author, or a group of authors, but consists of many individual contributions by different authors, this is called host document. In the second column of the list of references instead of an author's name there is the name of the editor with the note "(ed.)".

If there is no identifiable person as editor of the host document, but an institution (a corporate body), a short name of the institution with the location of the institution's head office in brackets appears as author in the second column. In the third column after the title there is the note "edited by <institution>", that means e.g. "edited by BBC (London)". If several institutions are involved you can solve this as follows: "edited by Siemens AG (Hannover); Nixdorf AG (Paderborn)" or "edited by NDR (Hannover) and University of Hannover".

Table 3.10 Popular citation styles in mathematics, natural sciences, engineering and social sciences (each institution have their own variation of typography)

Citation style name	Recommended by	Characteristics
Chicago Style	Chicago Manual of Style (CMOS)	Used by writers in many fields www.chicagomanualofstyle.org Sorting in articles: by order of citation Sorting in books: in alphabetical order
Harvard referencing	British Standards Institution, Modern Language Association, American Psychological Association	Parenthetical referencing (author-date system) Sorting: in alphabetical order
MHRA Style	Modern Humanities Research Association	Note referencing with bibliographic data in footnotes and list of references Sorting: by order of citation
ACS style	American Chemical Society	Parenthetical referencing (citation numbers) Sorting: in alphabetical order
AIP style	American Institute of Physics	Parenthetical referencing with literature numbers Sorting: alphabetically by author names
AMS style(s)	American Mathematical Society	Parenthetical referencing (author's initials and year, e.g. [AB90]), implemented in LaTeX Sorting: in alphabetical order
APA style	American Psychological Association Style	Parenthetical referencing (author-date system)
Vancouver system	Council of Science editors, American Society of Mechanical Engineers	Citation numbers in square brackets Sorting: in alphabetical order
IEEE style	Institution of Electrical and Electronics Engineers	Citation numbers in square brackets Sorting: by order of citation

Table 3.11 Author specification in the list of references

Number and description of authors	Entry in the list of references (examples)
– One author – Two or three authors – More than three authors – No author – Standards – Company publications – Institutional bodies	MILLER, K. SMITH, J.; SEBASTIAN, S. and KLING, M. MILLER, K. et al. (=Latin: and others) Dictionary of librarianship (=title words) ISO 4762 Bosch Rexroth Corp. in second column: BMVBS and in third column: ed.: Ministry for regional planning, building and urban development, commission for municipal gardens

Table 3.12 Usual abbreviations of terms in bibliographical data of publications

Bibliographical data	Abbreviations according to ISO 832
Book	bk.
Catalogue, catalog	cat.
Collaboration	collab.
Collection	coll.
Document	doc.
Editor, edition	ed.
Manuscript	ms.
Page	p.
Pages	pp.
Privately printed	priv. print.
Supplement	suppl.
Volume	vol.

When specifying authors, editors, and institutions as well as other information abbreviations may occur. Here the standards ISO 4 "Documentation – Rules for the abbreviation of title words and titles of publications" and ISO 832 "Documentation Bibliographic references – Abbreviations of typical words" give more information and rules, how words should be abbreviated. Table 3.12 shows a few examples.

The typography of the bibliographical data is partly optional and partly mandatory. It is optional, whether you want to write author's names always in small caps and titles in italic print. But a common system should be used consistently. For the punctuation there are common rules.

- At the end of the author name(s) and first names there is always a full stop.
- At the end of a title, there is always a full stop.
- After the location of the publisher, there is a space character, a colon, a space character and the name of the publisher.
- The word "publisher" is left out for well-known publishers. You may even leave out words from the name, if the publisher is still identifiable. Then you just write "Wiley" instead of "John Wiley & Sons". This item is closed with a comma.
- After the comma, you specify the edition and year. To specify the first edition is unusual. In this case, the year stands alone.
- After the year, there is a full stop.

If the information listed above is ambiguous, you may add notes to be more precise. Examples: Cambridge/UK or Cambridge/Mass., Available from NTIS: AD 683428.

Tables 3.13 and 3.14 show examples and the data structure of the bibliographical data for the most common publication types.

Here are examples for the data structure of bibliographical data of different publication types according to ISO 690 and example entries in a list of references.

General rule in ISO 690: First names can be abbreviated by initial letters. "Miller, William Thomas" will then be listed as "Miller, WT.". To distinguish the elements in the

Table 3.13 Examples for the bibliographical data of common publication types in Springer format

Publication type	Bibliographical data
Journal article (printed)	Smith J, Jones M Jr, Houghton L et al. (1999) Future of health insurance. N Engl J Med. 965:325–329
Journal article (online, only by DOI)	Slifka MK, Whitton JL (2000) Clinical implications of dysregulated cytokine production. J Mol Med. https://doi.org/10.1007/s001090000086
Journal article (online, no DOI available)	Marshall TG, Marshall FE (2003) New treatments emerge as sarcoidosis yields up its secrets. ClinMed NetPrints. http://clinmed.netprints.org/cgi/content/full/2003010001v1. Accessed 24 June 2004
Book (Monograph)	South J, Blass B (2001) The future of modern genomics. Blackwell, London
Book chapter (Contribution to host document)	Brown B, Aaron M (2001) The politics of nature. In: Smith J (ed) The rise of modern genomics. 3rd edn Wiley, New York

No commas between names and initials, no periods after initials or abbreviations

Table 3.14 Data structure for bibliographical data of common publication types according to ISO 690 and example entries

Publication type: Bibliographical data
Data in the data structure specifications given in italic print may be omitted
Journal article: Surname(s) and first name(s) of author(s). Title. Others involved (e.g. photographer). In: Title of the journal or publication. series, volume, etc., volume number (year of publication) issue number, pp. (first and last page number of contribution) McGUIRE, Gerald. Cyclotron design and efficiency. Journal on treatment of waste. 112 (1977) 9, pp. 12–20 WEAVER, William. The collectors: command performances. Architectural digest. 42 (1985) 12, p. 126–133
Book (Monograph): Surname(s) and first name(s) of author(s). (Edited by). Title. (series), (volume). Others involved (e.g. translator, editor). Location(s): publisher, edition (if not 1st edition), year of publication. If there is a CD-ROM, DVD, or video cassette as a supplement to the book, you have to add a note at the end of the bibliographical data, e.g. "incl. 1 DVD", ISBN-number LOMINIADZE, DG. Cyclotron waves in plasma. Translated by AN. Dellis; edited by SM. Hamberger. 1st ed. Oxford: Pergamon Press, 1981. 206 p. International series in natural philosophy. Translation of: Ciklotronnye volny v plazme. ISBN 0-08-021680
Contributions to host documents: Surname(s) and first name(s) of author(s). Title. In: Surname(s) and first name(s) of author(s). (Edited by). Title. (volume). Location(s): publisher, edition (if not 1st edition), year of publication. (section or chapter number and title, first and last page number of contribution), ISBN-number PARKER, TJ. and HASWELL, WD. A text-book of zoology. vol. 1, revised by WD. Lang. London: Macmillan, 5th ed., 1930. Section 12, Phylum Mollusca, p. 663–782

(continued)

Table 3.14 (continued)

Publication type: Bibliographical data
Contribution to a scientific series: Surname(s) and first name(s) of author(s). (Edited by). Title. In: Title of the series. (volume). Location(s): publisher, edition (if not 1st edition), year of publication. (other data, ISSN number). WRIGLEY, EA. Parish registers and the historian. In STEEL, DJ. National index of parish registers. London: Society of Genealogists, 1968, vol. 1, p. 155–167
Publications by companies, institutions, and other corporate bodies: Short name of the company or institution. Title. Edited by name of the editing company or institution (location of their head office). Publication number or similar data, Location(s): year of publication
Standards: Standard type and number (-part number):year of publication, Title. Location: publisher See reference part of this book
Intellectual property rights (e.g. patents, trademarks etc.): Surname(s) and first name(s) of author(s)/inventor(s) or company name. Title. Others involved. Note "Intellectual property right:", country code, document number, document type, date of publication. Eventual patent/trademark holder(s) CARL ZEISS JENA, VEB. Anordnung zur lichtelektrischen Erfassung der Mitte eines Lichtfeldes. Inventors: FEIST W, WAHNERT C, FEISTAUER E. Intellectual property right: Int. Cl.3: G 02 B 27/14. Schweiz, 608 626. Patent 1979-01-15 Larsen CE, Trip R, Johnson CR, inventors; Novoste Corporation, assignee. Methods for procedures related to the electrophysiology of the heart. US patent 5,529,067. 1995 Jun 25 European patent: EP 2013-B1 (1980-08-06) German patent: DE 27 51 782 1977-11-19
Electronic monograph, data base or computer program: Surname(s) and first name(s) of author(s). Title. Type of medium (database, monograph, computer program, bulletin board, electronic mail), storage of medium, (CD-ROM, online, magnetic tape, disk), others involved, edition (if not 1st edition). Location: publisher, year of publication, date of last modification/update or version, date when cited, eventually also time, other data like access data, ISBN/ISSN number CARROLL, Lewis. Alice's Adventures in Wonderland [online]. Texinfo. ed. 2.2. Dortmund (Germany): WindSpiel, November 1994 [cited 30 March 1995]. Chapter VII. A Mad Tea-Party. Available from World Wide Web: <http://www.germany.eu.net/books/carroll/alice_10. html#SEC13>
Contribution to a source of literature from the internet: Surname(s) and first name(s) of author(s). Title. In: series, eventually volume, issue number, first and last page number of contribution. DOI number or information regarding the last update and/or version, URL/URI or starting URL/URI and description of clicks you have to perform to go to the desired page, date when cited, eventually also time
Electronic mailing lists, discussion forums: Title. Type of medium, Location: publisher or owner of the forum, date when published, date of the citation, other data, access data Parker, Elliot. Re: Citing Electronic Journals. In Pacs-L (Public Access Computer Systems Forum) [online]. Houston (Tex.): University of Huston Libraries, 24 November 1989; 13:29:35 CST [cited 1 January 1995; 16:15 EST]. Available from Internet:

(continued)

Table 3.14 (continued)

Publication type: Bibliographical data
<telnet://brsuser@a.cni.org>
Additional data for personal messages and e-mails:
Title of the contribution or e-mail. Type of medium or location. Others involved/list of addressees/event, date of e-mail delivery or event

bibliographical data you should apply a consistent system of typography and punctuation, e.g. that every element is finished with a period. It does not matter, which system you apply. The main point is to use it consistently.

Other recommendations in ISO 690: A subtitle is distinguished from the title with a colon. Several publisher's locations are separated with a semicolon. The year or date of publication is specified as in ISO 2014 or as in the source of reference. Country, province and state names, which specify the names of cities more precisely, can be abbreviated according to ISO 3166. Title words can be abbreviated according to ISO 4.

Not all listed data is mandatory. If in doubt, please refer to the standard. One source for both parts is: http://openpdf.com/ebook/iso-690-1-pdf.html.

Now we want to give you some hints and information regarding special cases.

Special cases
If you cite from a brochure or manufacturer document and bind this document together with your Technical Report, please add a note in your list of references like "…, see Appendix C Other sources".

If sources of information are not accessible to the public, this must be clearly specified. You can add notes like "(in print)" or "oral statement/comment" or "citation from an e-mail written on 07.Jun.2005", "NDR radio transmission ,Auf ein Wort'on 12. Mar.2006" etc. In case of oral statements/comments it should be stated when exactly at which meeting in front of which audience the statement/comment was given (during a presentation, in a certain TV discussion etc.).

Additional information, which is useful for finding or buying literature (like ISBN for books, ISSN for journals, DOI for electronic documents, Library of Congress Control Number LCCN for documents registered by the American National Library, ordering number of a book etc.) can be added at the end of the bibliographical data.

If bibliographical data for articles in journals cannot be provided in the structure listed above, you should deliver similar information. Sometimes all required bibliographical data is printed on each page of the journal. However, quite frequently on the single pages of the journal the journal title is listed (often abbreviated) and the issue number, but the volume number by the publisher is missing. However, volume numbers are very important for finding journals in libraries, because at the end of the year the journals are bound as books. If there is only one volume per year, the libraries use the same volume number as the publishers and the volumes are identified by their volume number, e.g. "vol. 54". If there are several volumes per year and the page numbering runs through from the first issue to the last, the volumes are identified by their issue numbers or by page numbers. In this case the libraries use their own volume numbers. The bibliographical data of a single (fictive) article from volume 37 of the journal could then look like this:

Bibliographical data of an article in a journal

33. Liehr, J., Thermochemical Gasification of wood as useful removal of waste wood. In: Journal on treatment of waste. 37 (1985), pp. 824–836

Those who are looking for this article can look up the volume number by the publisher in the data base in the library. Using the data base is mandatory, because each library decides for themselves, into how many volumes the issues of one year of a journal are split.

If the issue number, year of publication and volume number are not printed onto the individual pages of the issues, the missing data may be found in the impress. The impress is often in the very front or back of an issue, in or near the table of contents. The journal volume numbers of the libraries can be found in the catalogues of the libraries.

Another problem is that journals sometimes have no issue numbers, but other issue labels. Then you should add these issue labels accordingly to the list of references, here Aug./Sept.

Specification of the issue, if there are no issue numbers

72 (1990), Aug./Sept., pp. 115–117

If a piece of information is published in a data network, it can be used by any other user in the network (provided the right access permissions are set). The users can save an electronic copy on your harddisk drive or a storage device, use and modify the piece of information as a whole or parts of it in any form.

These users could e.g. offend against copyright laws by deleting the author's name, using the information for commercial purposes without paying license fees, passing on the information to other users in the data network without prior asking for permission etc., and the temptation to do so is quite large. The rules in Table 3.15 help you to behave correctly.

These questions how to cite information from data networks correctly are relevant for any author of Technical Reports, who has access to such data networks (mainly the internet). If you have questions regarding complex texts or mathematical deductions you can ask for assistance in the web. If there are problems with the material collection, colleagues in the web may help. In principle, any information can be transferred via data networks. You as the author of a Technical Report should always specify "all used help and sources" truthfully to be on the safe side.

After the bibliographical data has been collected, we want to give you some hints regarding the typographic design of the list of references now. Lists of references are written with single spacing. However, if your supervisor always expects that a thesis must

Table 3.15 Dealing with information from data networks

1.	If the author or publisher of a piece of information formulates some limiting conditions, you should be fair and follow these conditions
2.	Citations from e-mails are treated as citations from physical letters or oral statements/comments (name the author and specify type and date of publication, do not manipulate the sense of the source text)
3.	Documents from the internet can be cited, as other sources of literature, literally or analogously. In both cases the bibliographical data must contain surname and first name(s) of the author, title. Either DOI number or date of last update or version, URL/URI or starting URL/URI and list of clicks you have to make to reach the desired page and the note "Accessed on: <date, eventually also time>" By the way: To limit the distribution of illegal contents, everybody is responsible for the information, he/she makes accessible via links
4.	If you wish to use information from the internet, you should think about whether you want to cite or link the information. If you want to cite it, download the information completely with all image files to your hard disk drive or another storage device. Note the URL/URI or the click sequence and the date, when you downloaded the information, because the operators of the homepage can change or delete the contents at any time. If you just want to link the information, the URL/URI and date of access is sufficient
5.	Computer programs can also be copied via data networks. You should obey the conditions for freeware/shareware (only complete, non-commercial transfer to third parties is allowed, license fees should be paid)

have 60 pages and you have only 55 pages, you can vary the spacing and deliver the required page number. Here is some flexibility. Since the transfer of the general rules to the individual sources of literature is hard for many authors, an example list of references shall facilitate this transfer, Table 3.16.

The document part "Literature" or "References" at the end of an article in a journal contains the same bibliographical data, but the layout is much more space-saving, Table 3.17. It depends on you, whether you want to apply the systematic and layout of the „space-saving form" (two columns) or the block format according to ISO 690. It is also up to you, to type author's names in capital letters or small caps and titles in italic print.

The rules of university institutes, companies, and other customers, i.e. the "rules of the house" must be followed. If you write a book or an article, the rules of the publishing company apply.

The correct setup of a list of references lasts much longer than estimated. Often missing bibliographical data must be collected. The strict rules regarding layout and sequence of information reduce the typing speed drastically. Therefore, you have to plan very large time reserves for writing the list of references. In case of doubt, the completeness of the bibliographical data is more important than following all the layout and sequence rules.

If you have to work with publications written in foreign languages that do not use Roman letters, the bibliographical data shall be transliterated/romanized in accordance

Table 3.16 Example of a three-column list of references

9 List of references		
1.	BOSCH	Global Responsibility—Environmental Report 2003/2004. www.bosch.com/content/language1/ downloads/UWB_en.pdf Accessed on: 04.Jun.2006
2.	YATES, JG.	Fundamentals of Fluidised Bed Chemical Processes. Butterworths Monographs in Chemical Engineering. London: Butterworths, 1983
3.	KUNII, D. and LEVENSPIEL, O.	Fluidisation Engineering. New York: Wiley, 1969
4.	DAVIDSON, JF., CLIFT, R. and HARRISON, D.	Fluidisation. London: Academic Press, 2nd ed. 1985
5.	ORMOZ, Z., CSUKAS, B. and PATAKI, K.	Studies on granulation in fluidised bed V.—Study on the particle size distribution of granulates. Hung J of Ind Chem Veszprem, Vol. 3 (1975) pp. 193–215
6.	WALDIE, B., WILKINSON, D. and ZACHRA, L.	Kinetics and mechanisms of growth in batch and continuous fluidised bed granulation. Chem Eng Sci, Vol 42 (1987), No. 4, pp. 653–655
7.	LAPPLE, CE., HENRY, JP. and BLAKE, DE.	Atomization—A survey and critique of the literature. Stanford research institute, Menlo, California, April 1976, available as microfiche AD 821 314 from British Library
8.	HAUSER, EA. et al.	The application of the high-speed motion picture camera to the research on the surface tension of liquids. J Phys Chem, Vol. 40 (1936), pp. 973–988
9.	KDG Flowmeters	Calibration chart for variable area glass tube size 14 and 14X, type E, F and G from KDG flowmeters, A division of KDG Instruments Ltd., Rotameter Works, 330 Puley Way, Croydon, Surrey, England, 1989
10.	HERING, H.	Powder Granule Generation in a Fluidised Bed (Pulvergranulaterzeugung in Wirbelschichten). Diploma thesis, Heriot-Watt-University, Department for Chemical and Process Engineering, Edingburgh/GB: 1989
11.	SKF	SKF Hauptkatalog (main catalogue). Catalogue 4000/IV T, Schweinfurt, 1994
12.	Wikipedia	Toluol. www.wikipedia.org/wiki/toluol, last update: 30. Apr.2006, accessed on: 04.Jun.2006
13.	Statistics Canada	Communications equipment manufacturers. Manufacturing and Primary Industries Division, Statistics Canada. Preliminary Edition.1970- Ottawa: Statistics Canada, 1971-. Annual census of manufacturers. Text in English and French. ISSN 0700-0758

(continued)

Table 3.16 (continued)

9 List of references		
14.	WEAVER, W.	The collectors: command performances. Photography by BRIGHT, RE. Architectural Digest. December 1985, Vol 42, no. 12, pp. 126–133
15.	HERING, L. and HERING, H.	How to write Technical Reports. Seminar documentation. 1998
16.	BÜRGI, F.	Scope and usage of multimedia in training and education. In: MELEZINEK, A. (ed.): The engineer in the joined Europe—Reflections and perspectives. 20 years IGIP. Presentations of the 21st International Symposium "Engineering Paedagogics '92". Leuchtturm-Schriftenreihe Vol. 30 (1992) Alsbach/Bergstraße: Leuchtturm-Verlag, pp. 221–226
17.	BS 188	Specifications for viscometers. 1977
18.	Yoo, J.J., Hinds, O., Ofen, N. et al.	When the brain is prepared to learn: Enhancing human learning using real-time fMRI. NeuroImage. https://doi.org/10.1016/j.neuroimage.2011.07.063

Table 3.17 Example of a space-saving list of references

5 References	
1.	BOSCH, Global Responsibility—Environmental Report 2003/2004. www.bosch.com/content/language1/downloads/UWB_en.pdf Accessed on: 04.Jun.2006
2.	Yates, JG. Fundamentals of Fluidised Bed Chemical Processes. Butterworths Monographs in Chemical Engineering. London: Butterworths, 1983
3.	Kunii, D. and Levenspiel, O. Fluidisation Engineering. New York: Wiley, 1969
4.	Davison, JF., Clift, R. and Harrison, D. Fluidisation. London: Academic Press, 2nd ed. 1985
5.	Ormoz, Z., Csukas, B. and Pataki, K. Studies on granulation in fluidised bed V.—Study on the particle size distribution of granulates. Hung J of Ind Chem Veszprem, Vol. 3 (1975), pp. 193–215

with the appropriate standard, e.g. "Medicinska akademija" or "Медицинска академия (Medicinska akademija)". The following ISO standards apply:

- ISO 9, Documentation—Transliteration of Slavic Cyrillic characters into Latin characters
- ISO 233, Documentation—Transliteration of Arabic characters into Latin characters
- ISO 259, Documentation—Transliteration of Hebrew characters into Latin characters
- ISO 843/R, Documentation—Transliteration of Greek characters into Latin characters
- ISO 7098, Documentation—Romanization of Chinese
- DIN 1460 Transliteration of Cyrillic characters (Slavic languages)
- DIN 31634 Transliteration of Greek characters

- DIN 31635 Transliteration of Arabic characters
- DIN 31636 Transliteration of Hebrew characters

If a title in the original language (e.g. the title of a book, an article or a journal) cannot be understood by the audience without doubt, it should be translated and the title in the original language added in brackets behind the English data.

3.6   The Text of the Technical Report

In Technical Reports (written or as a presentation) the technical terminology is written or spoken, taking the target group(s) into consideration. This technical terminology does not differ from the general language as much as it is often the case e.g. in psychological and sociological, medical and legal texts. The following sections provide you with hints, how to write good style in general and in Technical Reports, which peculiarities occur in the context of formulas and computations, and how you can improve the understandability of your texts.

3.6.1   Good Writing Style in General Texts

In schools as well as in universities reports, protocols and presentations are written with computers. This implies, that basic knowledge of structuring texts, typography, syntax, style etc. have already been taught there. Yet, in reality, this knowledge is in most cases improvable. The following overview shows general rules for a better understandability of texts, which are also valid for other non-fictional texts than Technical Reports.

General rules for a better understandability of texts

- Formulate short sentences.
- If possible, use only one main clause with one or two subordinate clauses or two main clauses connected with a comma or semicolon.
- Do not use too many foreign words.
- Explain foreign words and unknown abbreviations when they occur the first time.
- Avoid too many abbreviations.
- Use introductory and connecting sentences, to guide your readers with words. In these sentences you can:
    refer to the structure or table of contents,
    conclude the facts described so far,
    build a connection to the next document part or
    introduce a new document part.

- Use descriptive formulations, and write vividly and engagingly!
- Analogies, metaphors and comparisons create associations in readers' minds. Therefore, readers can recognize similarities with and deviations from their existing knowledge more easily.

One part of these general rules which improve the understandability of texts is the "iS", already introduced in Sect. 3.1.3. "iS" stands for "introductory sentence". You should avoid that after a document part heading there are figures, tables, or bullet lists directly following the heading without a connecting or introductory text. It is much better, if you use an "iS" at the beginning (and end) of each document part. Such introductory and connecting sentences tie in with the prior knowledge of your readers and structure the written information. All details are sorted under consideration of the "backbone". The readers are not left alone, but "guided with words". The frequent use of introductory and connecting sentences is a prerequisite that the readers understand your text without questions in the sense you want them to. Referring to the rule "Use descriptive formulations ..." here are three examples:

- **Analogy**: the orbit of an electron flying around the atom's nucleus is similar with the orbit of a planet travelling around its central star
- **Metaphor** (= imaginary expression in the figurative sense): a figure of the planet earth often stands for the internet
- **Comparison**: an electron in highly charged condition is like a tense spring.

If you are in doubt whether a formulation is understandable and corresponds to good writing style, you may wish to use the grammar checker of your word processor. Activate an option or preference or grammar rule. Your grammar checker will show you (possible) offenses against this rule by green wavy lies.

Now we want to deal with the rules, which should be applied especially for Technical Reports.

3.6.2 Good Writing Style in Technical Reports

The style of the Technical Report shall follow the general rules for good writing style above. However, there are also some rules, which are only valid for Technical Reports and which must be obeyed, too. Above all other rules, there is the following basic principle:

▶ Clarity and unambiguousness have always preference compared with writing style rules!

The Technical Report shall be understandable without questions for readers, who have technical knowledge but no detailed knowledge of the current project. Since authors of a Technical Report have often worked on their project for weeks or months, they can hardly imagine how much (or better: how little) a normal reader of the report can know about the project at all, when they start to write the report in the last phase of the project. Therefore, in Technical Reports too much detail knowledge is supposed very often, which the addressees do not have. This overstrains the readers of the report quite frequently, and their motivation to read the report is negatively influenced.

In addition, after frequent reading of own texts authors tend to become blinded by routine against their own formulations. Therefore, during the end check of the Technical Report you should show it a friend or colleague, so that he/she can proof whether the report is understandable for people who have not been involved in the project.

The Technical Report is usually written impersonally, i.e. passive sentences are used quite frequently and personal pronouns like "I, we, my, our, you, etc." are avoided. However, in a summary or critical appreciation it is OK, to speak of "we" or "our", if the own working group or department is meant.

It is traditionally like that. Most technicians got used to the impersonal way of writing during their education and professional practice. The customers will probably prefer impersonal writing as well, because they are used to it, too. For non-technicians passive sentences and impersonal writing seem clumsy, boring, and monotonous. Therefore, many books about good writing style recommend to write in active voice.

Passive sentences also comprise the danger, that the reader is uncertain who does something. For example, in a description of a machine or plant it may be uncertain, whether the operator or an automatic mechanism of the machine or plant executes an action. If this may occur, you should use an addition like "by the operating staff" (man) or "by the revolver control" (machine).

You have to decide carefully for your Technical Report, whether, how much and where you want to use active sentences instead of the usual passive. You are "on the safe side", if you avoid personal pronouns and use the passive voice instead. Here an example of formulating the same fact once in active and once in passive voice:

- **Active**: "… we have evaluated the following alternatives …"
- **Passive**: "… the following alternatives have been evaluated …"

However, using the personal pronoun "we" is bad writing style in Technical Reports!

The tense is present tense. Past tense is only used, if a previously used part, measuring procedure or similar is described.

The naming of technical appliances, assemblies, parts and procedures (e.g. in part lists and design descriptions) shall be according to the function of the appliance, assembly, part, or procedure, i.e. not handle but shutting handle, not gear but input gear, not plate but retaining plate, not angle but stiffening angle, not frame but carrying frame, not variant 1 but electric-mechanical solution etc. The general principle for naming parts is:

▶   Parts are always named according to their function.

This holds true in all areas of technology, e.g. also in civil engineering and electrical engineering. Now some examples as an illustration.

Company or working group internal names are probably unknown to the readers. Therefore, select neutral names.

Another aspect is, that you should use valid standardized terms for names of parts and processes as far as they exist. If you want to search for standardized terms in ISO standards go to www.iso.org and enter the word „vocabulary" plus a word from your field of technology. For example, entering the words „vocabulary engine" leads to ISO 2710-1:2000 „Reciprocating internal combustion engines – Vocabulary – Part 1: Terms for engine design and operation." To search the equivalent German standards, go to www. beuth.de and enter the search words „Begriffe Teil 1" plus a word from your field of technology. Example: „Begriffe Teil 1 Verbrennungsmotoren" also leads to ISO 2710-1.

If an assembly drawing and part list or a sketch of the described appliance belong to the Technical Report, it makes sense to add the position number in brackets after the part name. That is to say for example: "The fixation of the cover plate (23) was accomplished by groove pins to save costs." Here "23" is the position number and "cover plate" the name from the part list. This correlation of number and name in report and drawing as well as part list helps the reader, to find his way in the different documents.

Quite frequently, you can read the following or similar formulations in Technical Reports. "The design has a high mechanical strength and a very good wear resistance." In this case it is better, to substitute the general term "design" with the actual name of the complete (sub)assembly or plant. An application of this rule to the example above could be: "The oil mill has ...."

Often a value range is specified with two physical values. Then it is wrong to write: measure and measuring unit, extension sign (– or to) and again measure and measuring unit. Correct is to write the dimension only once, after the two measures. An example for illustration (text from the description of a mold for injection molding, wrong position of measuring unit is marked in boldface typing):

"The main problem is the effective insulation of the hot mold (ca. 160–200 °C) from the relatively cool distributor duct area (ca. 80–110 °C)."

▶   As a test you can read the written text loudly. Then you will find out fast, if
there are unnecessary details in the text.

To finish this section here is a further style rule for Technical Reports. Since the Technical Report is addressed to technicians, who normally approach all problems in a rational way, there should be no emotional and colloquial formulations in the Technical Report. Thus, sentences like "After the ergonomic analysis of his workplace the employee can continue to work joyfully." or "The programmed software ran really cool." should better be avoided.

3.6.3   Formulas and Computations

Formulas appear in Technical Reports mainly, if computations are developed. These computations can e.g. occur in the fields mathematics, physics and chemistry, but also in core areas of technology like civil engineering, mechanical engineering, electrical engineering, and informatics.

The formulas are often closely interwoven with the text, because the text refers to individual physical values, makes a statement on the formula etc. Text and formulas are often an inseparable unit.

In the technical education, pupils and students are confronted with teachers and professors using different formula signs for the same physical values in different subjects or lectures. This creates unnecessary disturbances and irritations. Here the standards try to help. Those who would like to go deeper into the fields formula symbols, formula syntax and formula layout can e.g. consult ISO 80000, Quantities and units with 15 parts covering the set of mathematical signs and symbols, the SI and the symbols for chemical elements:

– ISO 80000-1, General
– ISO 80000-2, Mathematical signs and symbols to be used in the natural sciences and technology
– ISO 80000-3, Space and time
– ISO 80000-4, Mechanics
– ISO 80000-5, Thermodynamics
– IEC 80000-6, Electromagnetism
– ISO 80000-7, Light
– ISO 80000-8, Acoustics
– ISO 80000-9, Physical chemistry and molecular physics
– ISO 80000-10, Atomic and nuclear physics
– ISO 80000-11, Characteristic numbers
– ISO 80000-12, Solid state physics
– IEC 80000-13, Information science and technology
– IEC 80000-14, Telebiometrics related to human physiology
– IEC 80000-15, Telebiometrics related to telehealth and world-wide telemedicines
– for German documents DIN 1301, 1302, 1303, 1304, 1313, 1338, 5473, and 5483

At first the term formula shall be defined. A formula can consist of formula symbols, physical values without dimension (constants), physical values with dimension (measures and measuring units, between them is exactly one space character) and mathematical or other operators, see Table 3.18.

For these formula symbols the information relevant for Technical Reports is now described. If a text contains only a small number of equations, the equations are emphasized by blank lines before and after the formula and eventually by an indentation of about 2 cm on DIN A4 paper. If a text contains many equations, the equations are

Table 3.18 Examples for constituents of formulas

Formula		
Formula symbols	Physical values	Mathematical and other operators
Vectors: a, b, c, ..., x, y, z or without boldface print with a right arrow above the vector Coordinates of vectors: $a^1, a^2, a^3, ...$ Scalars: a, b, c, ..., x, y, z Other formula symbols: matrices, column vectors, determinants etc. are displayed accordingly	Examples: length l in m time t in s velocity v in m/s mass m in kg force F in N pressure p in Pa power P in W concentration c in mmol/l quantity n in mol molar mass M in kg/mol	+, −, * or ·, : /or solidus $\pm = < \le > \ge \sim \approx \ll \gg$ $\times \perp \parallel \angle \nabla \cap \cup \supset \supseteq \not\subset \subset$ $\subseteq \in \notin$ etc. Besides: integral symbol $\int$ root symbol $\sqrt{x}$ exponents 23^6 indices L_{max} and diverse brackets

additionally numbered with equation numbers like "(16)" or "(3-16)" at the right margin of the printing area, so that you can better refer from the text to the equation. If a formula spreads across more than one line, the equation number appears in the last line of this formula.

▶ If you want to write formulas in HTML documents, you can save typing effort, if you visit the URL/URI www.mathe-online.at/formeln.

Comfortable word processors have a formula editor. If you want to type only a few simple formulas, you can insert fields or write the formulas as simple text.

In text mode, you should use the following settings: formula symbols italic, indices and exponents standard and superscript or subscript (=2 pt smaller than the basic font size); minus as operator for computations: dash (Alt + 0150), minus as algebraic sign: normal hyphen, multiplication dot: · (Alt + 0183). Example:

$$Q_{up} = c_w \cdot m_w \cdot (T_M - T_W^2) + c_{wk} \cdot (T_M - T_W^2)$$

More complex formulas can be typed in easier with a formula editor.

$$\Delta S = C_n \cdot m_n \cdot \ln\left(\frac{T_E}{T_{h,A}}\right) + c_t \cdot m_t\left(\frac{T_E}{T_{h,A}}\right)$$

▶ **Tricks for the formula editor:**

- Enter space characters with Ctrl + space bar.
- If ascenders and descenders are not printed completely (e.g. the upper line of the root symbol), select a slightly larger line spacing!

-   The proportional symbol can be typed in normal text and in the formula editor as AltGr-+.
-   The formatting templates Matrix/Vector and Text result in non-italic text (e.g. for measuring units).

If you have to write many formulas, you can also use the layout program "TEX" or the appertaining macro package "LATEX". It creates a very smooth and pretty formula layout. Some more image examples can be found in the Wikipedia article "Formula". Many publishing companies demand this format or offer at least layout templates for this format. The program is available as public domain software and widely spread in universities. An option to use the advantages of LATEX without having to install it on your computer is a Knoppix- or Kanopix-CD. You start your computer with the CD in the drive and the computer starts as Linux-PC. Depending on the distribution, you can then use LATEX directly, without having to install it. Ask someone in the next Linux club (or a "knowing" friend).

Formulas appear seldom as an individual formula, but mostly in more bulky computations with several formulas. The text should explain the formula symbols, describe formula transformations, pick up the results of the computations etc. It is desirable to name and explain the physical values, which appear in a formula, shortly before or after the formula in the text or in a legend.

Here is an example of fracture mechanics computations with explanations of the influencing factors in the text:

Text with formulas

IRWIN [28] developed another approach than GRIFFITH [23] to forecast the brittle fracture probability of ductile materials. IRWIN considered the elastic stress field ahead of the crack, which is characterized by the stress intensity factor K. With the stress σ and the crack length a it is:

$$K = \sigma \cdot \sqrt{\pi a} \cdot f\left(\frac{a}{W}\right) \qquad (3.1)$$

The factor $f(a/W)$ is dimensionless. It depends on the geometry of the specimen and the crack. Values for $f(a/W)$ are tabulated in standards, e.g. in BS 5447. Crack extension occurs, if the factor $\sqrt{\pi a}$ reaches a critical value, the so-called critical stress intensity factor

$$K_c = \sigma_f \cdot \sqrt{\pi a} \cdot f\left(\frac{a}{W}\right) = \sigma_f \cdot \sqrt{a} \cdot Y \qquad (3.2)$$

where σ_f is the fracture stress. The factor Y is introduced for simplification. It is computed as

$$Y = \sqrt{\pi} \cdot f\left(\frac{a}{W}\right) \tag{3.3}$$

The critical stress intensity factor K_c is obtained out of the experimental data very easily, since K_c^2 is given by the slope of a straight line of σ_f^2 against $1/a$, see Eq. 3.4.

The stress intensity factor refers to a specific load state. It is expressed via indices. I stands for normal stress, II for shear stress, perpendicular to the crack tip and III for shear stress parallel with the crack tip. The critical stress intensity factor for normal stress K_{Ic} is also called fracture toughness. If the plastically deformed zone at the crack tip is small compared with the geometry of the specimen, this value is independent of the specimen shape and size and therefore a material constant.

In such computations the formulas, the computed values and the text are connected with each other by signal words like "With <physical value> and <physical value> the result is <formula>." and "Taking into consideration <fact> and <formula> <formula value> can be computed as <formula>.".

According to ISO 7144 "Documentation – Presentation of theses and similar documents" formulas which are written in continuous text shall be written on one line (use e.g. $1/\sqrt{2}$ or $2^{-1/2}$ instead of solidus).

Important formulas, which play a central role for your Technical Report can also have a legend, where the influencing factors are explained. Here an example:

Formula with legend

$$m \cdot g = \sigma \cdot d_n \cdot \pi \cdot f\left(d_n / \sqrt[3]{V}\right) \tag{3.4}$$

m	mass of the drop and its satellites
g	accelearation of gravity
V	volume of the detached drop
d_n	orifice diameter
σ	surface tension of the liquid
$f(d_n / \sqrt[3]{V})$	empirical correction function, equals 1 for an ideal drop with no satellites and no material left on the tip.

Formula and legend form an optical unit, which the reader can find again easily when he reads the report the second time or later. All required explanations to the formula (including the citation) are listed in the legend. The reader can—but does not need to—read the connecting text. Typically in a legend of a formula the heading „legend" is missing.

From the text, you can refer to a formula by writing its number. The brackets are omitted: "..., see equation 4-5." Another example: "In equation 3-16 you can see ...."

Computations of loads differ from the computations described so far, because there is nearly no text. Computations of load have outcomes like "$C_{required} = 23.8$ kN" etc.

Then there must follow a statement like: "selected: grooved ball bearing 6305 with $C_{existing} = 25$ kN $> C_{required} = 23.8$ kN". Here is another example:

Computation of loads for the arms of the puller

Load assumption: All four pullers bear equal loads

$$\tau_S = \frac{F}{A} = \frac{100,000\,N}{50\,mm \cdot 40\,mm} = 50.00\,N/mm^2$$

$$\sigma_b = \frac{M}{W} = \frac{100,000\,N \cdot 35\,mm}{\frac{50 \cdot 40^2}{6}\,mm^3} = 262.50\,N/mm^2$$

$$\sigma_V = \sqrt{\sigma_B^2 + 3\tau_S^2} = \sqrt{262.50^2 + 3 \cdot 50^2}$$

selected: material S355JO

$$\Rightarrow \sigma_{allowed} = 355\,N/mm^2 > \sigma_{existing} = 276.42\,N/mm^2 \Rightarrow OK$$

In such computations, the text stands back very much. There are only a few connecting words. The information is expressed nearly only in formulas and equations. This writing style may only be used in computations, which aim at finding numerical data.

All required formulas must be explicitly written, then the measures should be inserted —where useful with measuring unit—and in the end there follows the result of the computation. This procedure is compact and reproducible. For a better overview in case of formulas, which spread across more than one line, you should put the first equation sign of each line below the previous one.

For general discussions, developments etc. the writing style in the fracture mechanics example above (a few formulas, a lot of text) must be used.

3.6.4   Understandable Writing in Technical Reports

Understandability is a complex term in communication sciences. Whether the target group understands a text depends in general on two groups of properties. These are the "text properties" and the "reader properties".

Reader properties cannot be influenced by the author, since they depend only on the planned or random readers of the text. They are:

– general prior knowledge of the readers,
– their ability to concentrate on reading and their routine in reading,
– their motivation for or inner aversion against the topic(s) of the text,
– their command of the language in which the text is written,
– their knowledge of technical terms, and
– their expertise in the described field of knowledge.

The text properties are specified by the author beside a few exceptions like layout rules according to a corporate identity. Typical checklist questions referring to text properties are:

– Is the title descriptive and does it create interest?
– Is there a structure and is it logical ("backbone")?
– Are paragraphs too long?
– Are sentences too long?
– Are there nested sentences?
– Are unknown words explained when they are used for the first time?
– Are unknown abbreviations explained when they are used for the first time?
– Are there too many foreign words?
– Has the reader all information available in all sections of the Technical Report, which he/she needs for following the contents?

This bullet list shows clearly, that the understandability of the text can be negatively influenced by an inappropriate design of the text properties above. Therefore, we want to describe some possibilities to improve the understandability of non-fictional texts.

Improvements of the understandability of texts can be made on three levels. These are the text level, sentence level and word level, which are individually looked at in the following.

Improvement of the understandability on text level
On the text level the elements that improve the understandability are at first the indices and lists, i.e. table of contents, index, glossary and lists of figures, tables, abbreviations, units and symbols etc.

In addition, headers and footers, marginalia as well as page numbers and cross-references improve the understandability on text level.

Moreover, introductory sentences at the beginning of a document part and connecting sentences at the end of a document part have to be mentioned here. These sentences can also be used for the introduction of figures and tables. They help the reader, to follow the "backbone", because they rebuild a connection to the logical structure again. They describe, which information follows next and why, how the following information must be evaluated, which significance the information described so far has etc.

Another measure, which improves the understandability of the text are footnotes. Footnotes are superordinated numbers, which are consecutively numbered in the text. By default, the number appears behind the text it refers to and it is repeated at the bottom of the page below a left justified, horizontal line of about 4–5 cm length. There you can place information, which would disturb the normal flow of reading. Examples for such information are

- comments about cited texts,
- bibliographical data of cited literature,
- the exact location of the cited information in the source of literature and
- hints or comments regarding the normal, continuous text.

In most cases (as in ISO 7144 and several other ISO standards) the footnotes appear in a smaller font size than the standard text. However, DIN 5008 recommends to write them in the same font size as the standard text. Footnotes are used quite frequently in the humanities.[1] In the technical sciences, footnotes are less common.

If one or more footnotes refer to the contents of a table, they appear directly below the table without a horizontal line of 4–5 cm. The German Association of Electrical Engineers recommends that normal footnotes get superscript numbers and tables get superscript small letters to distinguish them from one another. You can manually create the superscript letters (Format—Character, Superscript).

For shorter texts, sometimes endnotes are used, beginning on the last page of the text. Endnotes can be converted to footnotes by pressing a button and vice versa.

The area of typography (page margins, font type, font size, etc.) brings also possibilities to improve the understandability of non-fictional texts, but only in a wider sense. In a wider sense, because the typography belongs more to recognizability or perceptibility than to understandability.

A quite important rule in this context is: The wider the printing area is (or the longer the lines are), the larger the line spacing should be.

(Good) figures and tables open up the text, complete it and address other apperception channels than continuous text. So, the author can display the information in several ways and on more than one level and the reader can follow the flow of information better. Therefore you should use many figures, tables and bullet lists as well as pictorial or tabular re-arrangements of text, see Sects. 3.3.4 and 3.4.8. Examples also make the text illustrative. However, too many examples can have a negative influence on the briefness and conciseness of your texts.

Improvement of the understandability on sentence level The following overview shows some rules which improve the understandability of non-fictional texts very much. In addition, the translatability into foreign languages is improved, if you keep these rules.

> **Rules for better understandability of texts on word level and sentence level**
>
> - The sentences should be as short and simple as possible.
> - Each new fact should preferably be described in a new sentence.

---

[1]Here is an example of a footnote, like it is usual in the humanities.

- Leaving out verbs to shorten sentences is not allowed.
- The sentences should not be longer than 25–30 words.
- Paragraphs should have maximum six sentences. Paragraphs with only one sentence should not appear too often.
- Tables and bullet lists should be used as often as appropriate.
- Compound tenses should be avoided (depending on the target group), the simple tenses (present tense, past tense and future tense I) are better understandable for most people.
- Abstract nouns (foreign words) are tiring for the readers and should therefore be avoided.
- Meaningless phrases and filler words also appear as tiring, if they are used too often, and should be used carefully.
- If a word is used in an unusual meaning, please use quotation marks or italic printing.
- The first verb should not appear too far at the end of the sentence.
- Nested sentences should be avoided. Parentheses in such sentences should be only short. Their contents can eventually be expressed in individual sentences.
- Double negations are in most cases superfluous. A normal negation shall not appear too far at the end of a sentence.

The sentence structure is very important for the understandability of the text. In general one main clause plus one or (more seldom) two sub clauses or a combination of two main clauses should not be exceeded.

The understandability of the text on sentence level can be improved very much by using conjunctions and prepositions. By using conjunctions you design a logical structure of the individual sentence parts and connect the current sentence congruously to the previous one. Here are some examples:

Long sentence without conjunctions (with improvement)

not so easy to understand:
Cranes always have a maximum tolerable lifting load, which when excessed, can result in a buckled crane boom, an overturned crane or broken lifting ropes.

better understandable by conjunctions:
Cranes always have a maximum tolerable lifting load. **If** it is considerably exceeded, **then** the crane boom can buckle, the crane can overturn, or the lifting ropes can break.

A list of conjunctions can be found here: http://grammar.yourdictionary.com/parts-of-speech/conjunctions/conjunctions.html.

Use clear references!

Misleading reference to previous sentence (with improvement)

not so easy to understand:
The current I flows through the resistance R which is relatively small.
Who is meant here? I or R?

better understandable due to clear references:
The current I flows through the resistance R. R is relatively small.

Use clear, precise and specific formulations. Write „white" or „black", not „gray".

Unclear wording (with improvement)

not so easy to understand:
The new sensor was much more linear than the old one.

better understandable due to precise expressions and facts:
The sensor characteristic of the new sensor has a deviation from a linear relationship between current and flow rate of max. 3%. In case of the old sensor the deviation was max. 7.6%.

Expressions like rather complex, nearly linear, very fast, little power, highly sensitive and relatively low are much too vague and must be supported with figures, if they are important. If they are unimportant, they should be left out.

▶   Try out to translate your text to a foreign language in your mind. If you do so, you will clearly recognize your complex sentence structures, unclear references and imprecise descriptions. You will probably wonder, how many alternative and more simple expressions and sentence structures you find, while you translate.

Improvement of the understandability on word level
On word level the amount of simple, common, accurate words shall be as high as possible. Technical terms and abbreviations shall be explained when they are used for the first time. In addition, technical terms and abbreviations can be explained in a glossary, abbreviations eventually also in a separate list of abbreviations.

In general the number of used foreign words shall be limited. Technical terms sometimes consist of several nouns. In this case, you should use hyphens between words which belong together.

The following rule plays a central role for the understandability of text on word level:

▶   Within one Technical Report the same appliance, assembly, part, fact or pro-
      cess is always consistently called by the same name, otherwise the reader is
      unnecessarily irritated!

This rule also applies, if within one paragraph one part is addressed all the time. In the
language courses, they emphasize fluent speech. Sentence introductions, verbs and nouns
are substituted by synonyms, so that the text does not become boring. In your word
processor, you can mark a word and look up synonyms with the thesaurus.

However, a Technical Report—other than a lyrical text—has to transport information
without any ambiguity. Technical parts, facts and procedures may therefore be addressed
only with their once specified names. Otherwise misunderstandings and mistakes can
occur—with potentially severe consequences.

It is important that you do not presume too much prior knowledge. Authors overes-
timate the knowledge of their readers over and over again. Apply all suitable measures to
make your texts clear, understandable and easy to read (figures, tables, bullet lists,
intermediate headings etc.).

3.7   Using Word Processing and Desktop Publishing (DTP) Systems

Today the usage of word processing systems is state-of-the-art. Since texts may be edited
easily, some authors tend to write "something" to begin with. If at the end, when the
project deadline is in sight, the time-pressure rises, necessary changes remain undone. If
you look at Technical Reports you may see, from when on the time-pressure became very
hard. Problems with inner logic, spelling, grammar as well as creating or integrating
figures occur much more often from then on and the report creates the impression of being
made "quick and dirty". To avoid this, you should have enough time for proof-reading,
entering the corrections and for the "end-check" in your project plan.

When you define the page layout and the formatting templates, you have to keep the
layout rules of your university or company regarding their corporate design. If there is a
lack of rules in the corporate design guidelines, you may use the rules we propose here
accordingly.

The definition of the page margins and the placement of the page numbers are flexible
to a certain extent. The following rules have proven to be practical: Define the page
margins right from the beginning of writing your Technical Report, so that the printing
area is fixed and the line and page break do not change very much shortly before the
deadline of your project.

This includes, that in a group all "writing" members use the same settings and formatting templates, e.g. for page margins, header, document part headings etc. and have the
same fonts installed.

Ideally you should speak with your copy shop or printer already at the beginning of your project, whether you provide the printing data in digital form or whether paper originals shall be copied. For digital printing, you often need a special printer driver. You should install it already at the beginning, so that the line and page break do not change much any more shortly before the deadline of your project.

If the front and back sides of the pages of the copied Technical Report shall be used, you have to switch on alternating page margins in your word processor, see File or Format menu or Page layout tab—Page setup or Page margins.

The upper and lower margin should be at least 20 mm wide on DIN A4 paper, if there is no header or page number. There should be a gap between header or footer or page number and the edge of the paper of at least 15 mm. The right/outer margin may not be narrower than 15 mm, 20 mm is better. The left/inner margin should be at least 25 mm wide in any case so that the text is visible also at the beginning of the lines, Fig. 3.28.

When you define the page margins, you have to consider how the Technical Report will be bound later. If you use staples, plastic or cardboard folders, plastic spiral (comb) binding, wire-o-binding (with wire spirals) and staple binding you need a wider inner margin compared with an adhesive binding. Also, if the size of the pages of the report is to be reduced with the copier (e.g. to DIN A5), the margins on the DIN A4 original must be wider.

When entering the page numbers you have to distinguish at first between book page numbering and report page numbering. In books, the front and back side of the page are used, while in Technical Reports usually only the front side of the pages is used. The page

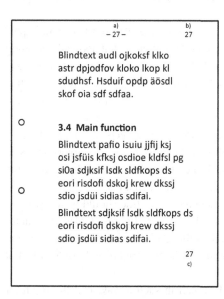 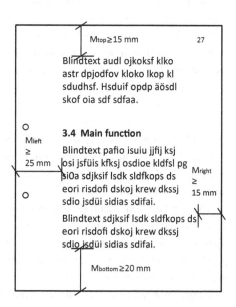

Fig. 3.28 Page layout with usual positions of the page number (positions a and b are recommended) and minimum size of the page margins

numbers can be placed in the middle of the top margin, on the right side of the top margin, or on the right side of the bottom margin.

If the right and left pages shall be used, the page number must always appear on the outer edge of the pages and you need just and injust headers. On the right side there are always the odd page numbers (1, 3, 5, 7 etc.).

The design of headers and footers shall be decent. Headers which expand across more than one line—eventually even with a logo—often appear to be overloaded. In the header, you can show the chapter heading and the page number. A thin line (underlining), an appropriate distance to the normal text or a different font style (e.g. italic) as well as a smaller font size can be used to distinguish the header and footer from the ordinary text. If you use a thin line, it should start at the left edge of the printing area and end at the right edge of the printing area, i.e. the thin line is exactly as long as the printing area is wide.

If the front and back sides of the pages shall be used and in the header there is the chapter heading on the left and the subchapter heading on the right page, this is called column titles. The headings in this book are an example for these column titles.

In the footer—also divided from the normal text by a thin line—you can give information about the version of the document and/or a copyright note.

The example pages above show, how and where page numbers can be placed and how the margins can be defined.

Up to now we have introduced recommendations for the margins and headers and footers. Now the text in the printing area shall be structured and designed by typographic means. Such settings refer to a paragraph. Therefore, they call it paragraph layout. The paragraph layout can be controlled by the formatting functions or by applying the paragraph formatting templates.

There should be a vertical spacing of ≥ 6 pt up to one line between two paragraphs.

For document part headings, you should take as a rule of thumb, that above the heading there should be two blank lines and one blank line below it. In any case the distance above the document part heading shall be distinctively larger than the distance below it, so that the reader can recognize more easily, that a new document part starts here and which document part the heading belongs to, see example pages above. For figure subheadings and table headings this rule should be applied accordingly, see Sects. 3.3.1 and 3.4.2.

Document part headings are printed in boldface in most cases. A larger font size compared with the normal text is recommended. Depending on the hierarchy level of the document part heading there are different font sizes for the headings.

The gap between document part number and document part title should be at least two space characters wide. If you create the tables (of contents, figures, tables etc.) automatically, this gap should be created by a tab. Then the tab is copied from the document part heading in the text to the automatically created table of contents and you can layout the table of contents more easily.

Figure subheadings and table headings are layouted in such a way, that the figure or table title can be read and their labels can be distinguished from the normal text quite easily, because in cross-references to figures or tables in the normal text these labels are used as search criteria. If you look for the figure/table to which the cross-reference refers, the information: "here is a figure/table" comes out clearly. To identify the figure or table number more easily, bold print of the label (like "Figure 12" or "Table 17") within the figure subheading or table heading has proven to be practical. The label is distinguished from the title of the figure subheading or table heading by two space characters or—better —by a tab. A colon behind the number is no longer usual. Here are an example for a figure subheading and an example for a table heading.

Figure 12 Principle of under-powder welding

Table 17 Filter weight depending on the welding position

Figure subheadings or table headings which spread across more than one line are layouted so that all lines of the figure or table title start at a common building line (use tab or hanging indentation!).

The line spacing "1 ½" is a good value for Technical Reports. Today they also use other (smaller) values for the line spacing. However, the eye can hold the line better, if the spacing is 1 ½. Very bulky works can become too thick and unhandy with this large line spacing. To avoid this, you may use both sides of the pages for printing or apply a smaller line spacing.

Now there are some remarks regarding the text justification. Normal text is either printed left justified or—more often—left- and right justified in the Technical Report. The left- and right-justification emphasizes vertical lines. This strengthens the optical effect of indentations and building lines. Moreover, it looks more pretty. Figure subheadings and table headings may be centered, if the figures and tables are also horizontally centered. The title of a short article may also be centered, if there is no title leaf.

Indentations may be created in different ways. For example, the following operations result in an indentation:

– inserting space characters (however, this does not result in a defined indentation in case of left- and right-justification!),
– defining left justified tabs,
– moving the left indentation markers in the current paragraph (hanging indentation),
– defining an indentation via menu,
– using a table without borders and leaving the first column empty.

In practice, there are often texts which use many different options to create an indentation within one document and which have very different indentation values. The result is a layout, which appears untidy, and when you enter text corrections, you have to find out first, with which method the indentations have been created. Therefore, you should use only a few methods of creating indentations and only a few indentation values (e.g. tabs at 5, 10, 15, 20 and 25 mm).

List structures can be automatically created with the two icons Numbered list and Bullet list. Nested lists can be created with further indentation.

The selection of the font type is also a very important decision. Fonts that are similar to Times have proven to be adequate for larger amounts of text. Due to the serifs (the small lines at the ends of the letters), the reader can hold the line well during reading. When he switches to the next line, his eyes do not erroneously jump to the over next line. The reader is well accustomed to these fonts from reading newspapers or books. Fonts without serifs (e.g. Arial, Helvetica) can be used for title leaves and overhead slides.

The normal fonts are proportional fonts. The distance from the end of one letter to the beginning of the next letter is constant. These fonts are not so well suited for the design of figure tables, where it is necessary, that the figures have the single, tenner, hundred and thousand digits aligned one below the other. For example, the digit "1" is narrower than the other digits. A font with fixed spacing (e.g. Courier, Lucida Console, Monospaced) might help in this case. Here the distance from the middle of one letter to the middle of the next letter is constant. If you design tables with these fonts, you will not have to use tabulators so much. Simply adding space characters is enough to create common building lines for tables, bullet lists, equations etc. The tables and bullet lists are then easier to read. This is especially the case for long tables and computer listings.

For the font size the following values should be applied. The usual fonts from the Times family should not be smaller than 9 pt. This is the lower limit of readability. A font size of 10, 11 or 12 pt is well readable. The standard font size should not be larger. If the size of the Technical Report shall be reduced to DIN A5 with a photocopier, the font size should be 13 or 14 pt for the standard text.

Text accentuations can be achieved in different ways: bold, italic, underlined, bold and italic, small caps, capital letters (majuscules), but also by frames or a fill.

These accentuations have different functions. They can show literature citations [MILLER, 1989, p. 151], emphasize **important** parts of the text or highlight an <u>un</u>usual sense. Also, "technical terms" may be denoted, or via (an annotation or) the insertion—idea—of thoughts words or text passages may be accentuated. However, the text accentuation also disturbs the flow of reading. Depending on the text type, you should apply a different amount of text accentuations:

– The classic book layout allows only little text accentuations. If they cannot be avoided, in seldom cases italic print is allowed.
– Advertising texts are the other extreme. They are created to get and hold the attention of the reader and to position the message "buy me" as deep as possible in the readers heart and mind. They address the human psyche in a clever way and apply unusual effects of text and image design, create new words, and exceptional sentence structures to achieve more turnover. Everything is allowed, which is successful and not displeasing.
– Instructional texts (user, maintenance, repair, installation or operation manuals, training and seminar documentations etc.) use text accentuations for attracting and controlling the readers' attention (like this book, which is therefore a little different from the classic book layout). Text accentuations use the reading conventions of the people to display information so prominently that they cannot be overlooked.

However, too many text accentuations are too much of a good thing. If too many information pieces are marked as important by means of text accentuations, the reader cannot decide any more, which information is really the most important one. He/she is irritated and frustrated.

You as the author of a Technical Report have to find an acceptable balance. You have to keep in mind your supervisor or customer and the target group. The central question is: How much creativity and unconventional design is acceptable?

– Are the supervisor or customer and the target group conservative? Then select a simple and decent layout! Technical information is assumed to be much more important than a functional and clear design.
– Are the supervisor or customer and the target group informal and unconventional? Then the text may be formulated informally and more text accentuations are allowed!
– Are the supervisor or customer and the target group top managers? Then you should do anything you can to shorten the text and present important information in graphics.

Moreover, it is important, how the text is probably read. Is it read like a textbook sequentially or like an instruction manual or an encyclopedia selectively? If you assume, that the text is read in a selective way, it is extremely important to work with many text accentuations. Beside the already mentioned text accentuations with different printing (bold, italic etc.), there are the following options:

– **Marginal notes**
   At the outer margin of the document in an own column there are keywords which represent the contents of the appertaining paragraph. Well-suited for introductory non-fictional texts.
– **Register**
   At the outer edge of the pages there are black or colored marks. This helps to distinguish different chapters. Well-suited for encyclopedias and product catalogues.
– **Column titles**
   Well-suited for alphabetically-ordered information. At the upper margin of the document on the left side there is the first and eventually on the right side the last alphabetical entry on the current page. Example: Telephone book or encyclopedia.

The following questions are essential for the application of all text accentuations: Which information must be found how fast and secure? Which search strategy do the readers apply? Which reading conventions can be assumed in the target group? Moreover, even a very sophisticated system of attracting and controlling the readers' attention may not forget the following basic condition:

▶   To achieve that the readers can use the usual lists and tables and marginal notes at all (table of contents, index, list of figures, list of tables etc.), they must know the words being used there and search for them themselves! This means that these entries must be answers to questions asked by the readers (not by the author).

Now there are some rules for a good page make-up. There are some information units, which must never be separated by a page break.

- A document part heading may never stand alone or together with only one text line at the bottom of a page.
- A single text line of a paragraph may stand alone neither at the bottom nor at the top of a page, unless the paragraph consists of only one line.
- A table heading and the appertaining table may not be separated from each other by a page break. In addition, the individual rows of a table may not just be split. In this case, the table must be continued on the next page with the same table heading and a continuation note. Alternatively, you can move a paragraph above the table, which stood below the table or vice versa.
- This is also true for figures and figure subheadings.

If possible, figures shall be located on the same page as the appertaining text or on the following page, so that the information in the figure is presented near the text. This improves the text-figure-relationship and the understandability of the Technical Report. If you want to avoid that the figure is positioned on the next page, you can only shorten the text, move a paragraph below the figure, use smaller empty lines (e.g. 10 pt instead of 12 pt) or apply the function Format—Paragraph—Distance below (a paragraph).

At the end of a chapter, the last page before the next chapter heading is quite frequently not completely filled. If in such a case on the last page of the preceding document part the page is at least filled by 1/3, this is acceptable. Otherwise some paragraphs can be moved from one file to the next or the text may be shortened or the size of figures may be enlarged or reduced, so that the page break fits better.

It is clearer, if each chapter begins on a new page. If both sides of a document are printed, sometimes a new chapter may only start on the right side where the pages have odd page numbers. In this case, the last page of the preceding chapter—on the left side—may remain empty. In the Technical Report, in most cases the report page numbering is applied and the back sides of the pages are not printed. In this case, it is best, if each chapter begins on a new page.

The line break may also be influenced systematically. The term line break contains all measures of an author, which result in a new line. In document part headings, figure subheadings and table headings it may be undesired that a word is automatically hyphenated. Then you can create a line break with the key combination Shift + Enter. Also in bullet lists, it may make sense, to enforce a new line in this way, so that the reader can read the information in specific information blocks controlled by the author.

If the text is left justified and right justified, you have to enter a tab before the enforced line break, so that the last line before the line break is left justified and large gaps between words are avoided. If you switch on the display of non-printable characters (¶) this looks like: → ↵.

In the continuous text the line break can mainly be influenced by inserting hyphenation proposals (optional hyphen, soft hyphen), if the automatic hyphenation results in undesired or wrong hyphenation, or if due to missing automatic hyphenation there are very large gaps between the words (in extreme cases even between the individual characters). To influence the hyphenation, there are special characters. These are non-breaking space, protected hyphen and soft hyphen.

The non-breaking space is used between the components of a multi-part abbreviation or between abbreviated academic titles and family names. It creates a constant word distance, which may be smaller than between normal word distances, if the text is left justified and right justified. Also the non-breaking space prevents the automatic hypenation, so that names like Dr. Mayworth, figure or table labels like Figure 26 as well as value and physical unit like 30 °C, 475 MPa etc. are not divided, but stand together on the next line.

The protected hyphen shall—similar to the non-breaking space (or protected space)—circumvent a hyphenation. It is e.g. used in words combined of a letter and a word like e-mail, X-rays etc.

If you enter hyphens, you should insert soft hyphens. If you enter normal (hard) hyphens (Minus on the normal keyboard) and insert or delete text later on, the hyphens appear somewhere in the middle of the line, which should be corrected, but which is often forgotten or overlooked. In this context look carefully at newspapers and journals. There this error also occurs quite frequently.

Sometimes there are terms consisting of combined words, which can be connected with a hyphen. This improves the readability and understandability of your text. Please use such hyphenation consistently throughout the whole Technical Report. Example: "rubber block drive belt" becomes "rubber-block drive-belt".

While editing text, you are sometimes astonished about automatic functions. They shall help to avoid frequent mistakes, but sometimes they are very annoying. If such an automatic function creates undesired results, you should look at the AutoCorrect Options of your word processor. By the way, in the AutoCorrection table you can delete undesired entries and enter your own, typical typing errors, if they shall be automatically corrected. Also quite interesting are the settings regarding "Compatibility" and "Spelling and Grammar". You should read carefully there. In most cases, there is an option, which controls the undesired effects, which can be switched to try out working with different settings than the standard settings.

It is generally recommended, to format larger documents with formatting styles instead of individual assignments for each paragraph. This is true for all frequently used layout patterns like document part and table headings, figure subheadings, bullet lists, cited text etc. Indentations should not be created with space characters, but with tabs or indentations. Such an approach saves work and assures a uniform layout. Automatic cross-references also save work.

3.8 Completion of the Technical Report

The phase "completion of the Technical Report" consists of the tasks proof-reading, entering the corrections, creating the master printouts, end-check as well as copying, binding and distributing the report or creating a PDF file from it and publish it in a data network. No later than directly before the last proofreading you should talk with your copy shop or print shop, whether you shall provide digital data or printed master copies. For digital printing, it may be necessary, that you use a special printer driver. This often influences the line and page break.

3.8.1 The Report Checklist Assures Quality and Completeness

In this phase there is mostly a lot of time-pressure. To find as many errors as possible despite the time-pressure, we have collected our experiences in the following report checklist, which shall serve you as a companion during all tasks in the phase "completion of the Technical Report". If you check all points in your Technical Report which are contained in the list and correct them as necessary, there is a high probability that the most frequent mistakes are eliminated. Therefore, we recommend, that you check your Technical Report with this report checklist. The quality, which can be assured with this approach, can hardly be achieved without using the report checklist.

Report checklist
Table of Contents:

- Heading "Contents" (according to ISO 2145)?
- Are there page numbers? is only the first page of every document part listed?
- Are page numbers aligned right justified?
- Document part number, heading and page number must be identical in the text and in the table of contents!
- Has the automatically created table of contents been updated?

Contents:

- The names of sub functions must be identical in the verbal assessment, in the morphological box and in all other descriptions in the Technical Report regarding number, name and order.
- Reduce sentences in your mind to check, whether they are logic, i.e. leave out sub clauses and ellipses. Is there a complete sentence left?

Spelling and layout:

- Is the text left- and right justified?
- Is automatic hyphenation switched on?
- Is the sign for a range of numerical values either "to" or "–"?

- No space character before full stop, colon, comma, semicolon, exclamation mark.
- Always use a tab between bullet and list item.
- Are bullets used consistently on different hierarchy levels in lists?
- There is always a space character between physical value and physical unit.
- Before and after each mathematical operator (+, −, ×, :, <, >, = etc.) there must be one space character. The signs + and - are placed before the numerical value without space character.
- All types of brackets, i.e. (...), [...], /.../, {...} and <...> enclose the text directly without space characters. The same is true for quotes "" and ''.

Figures and tables:

- Figures have a figure subheading, tables a table heading.
- Avoid too thin lines (line thickness ¼ pt) because of problems when copying.
- Figure and table titles always begin with a capital letter.
- Adjectives in tables and lists of requirements should consistently begin with small letters (possible exceptions: table header and introductory column).

Technical drawings:

- Is the title block filled in completely?
- Single part drawings: Can the part be manufactured? Are all measures available which are needed for manufacturing?
- Assembly drawings: Can all parts be mounted? Are all required assembly dimensions given?

Literature work:

- Bibliographical data for cited literature and information where to find the literature in the library should be noted on the first page of the copied sheets directly after you have copied the pages.
- Check these data again before you give back the literature to the library.
- Check bibliographical data in the list of references for completeness.

Line breaks:

- Are the results of the automatic hyphenation correct?
- If there are too wide gaps between words (when text is left- and right justified) or if the line width differs too much (when text is left justified) insert hyphenation proposals (Ctrl-Minus).
- If possible, after hyphenation there should not be one syllable alone in a line.

Page breaks:

- Document part heading alone at the bottom of a page and appertaining text on the next page is a no-go!

- One line standing alone at the end or beginning of a page should be avoided.
- Landscape format must be readable, if you turn the page by 90° clockwise (binding margin is at the top).

Miscellaneous:

- Cross references in the text referring to document part numbers, figure and table numbers, page numbers, document part headings etc. should be checked for correct numbers and compared with the table of contents, list of figures, list of tables, etc.
- Folded tables and figures (swing-outs) should be folded so that they are not cut into pieces, when the report is cut after binding.

Completion of the Technical Report:

- Plan sufficient time for proofreading and end-check!
- If you use formatting templates and automatic creation of lists and cross-references, all fields must be updated.
- When proofreading, check the text of cross-references to figures and tables, whether you refer to the right figure or table resp.
- Pages, which must be copied with photo key, should be processed separately.
- Figures, which need to be glued in, must be cut so that they have straight and rectangular edges. They should be glued onto the copy original pages with drawing board and ruler.
- Always use a new binder.

3.8.2   Proof-Reading and Text Correction According to ISO 5776

During the final proofreading, see Fig. 3.29, all remaining errors in the Technical Report should be found. The following error types are nearly always still existing in the Technical Report: wrong wording or wrong vocabulary (if the report is written in a foreign language), spelling errors, wrong commas, grammar mistakes, bad writing style, repetition of contents, errors in the logical sequence of thoughts, layout errors and miscellaneous errors (readability, bad briefing of the typist).

If you can concentrate well, search all errors in one proofreading cycle. Otherwise group the types of errors, you want to concentrate on, into groups and search the errors in several proofreading cycles. You can find all error types far better, if you do the proof-reading on paper printouts rather than on the screen. On paper, the contours of the letters have more contrast and sharpness. In addition, on a sheet of paper you can look at far

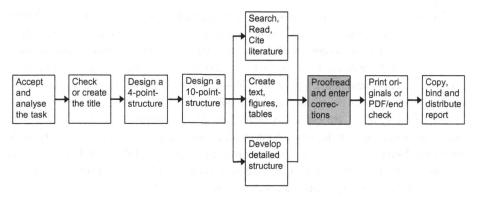

Fig. 3.29 Network plan for the creation of Technical Reports: proofreading

more information in a context, as on a screen and turning the pages back and forth is much easier. This facilitates looking for repetitions of contents a lot. Enter the corrections, which you marked on the paper very soon into the appropriate text files!

While entering the corrections and reading the text on the screen, you will often find additional open problems. Try to become used to standardized personal notes and symbols for those things, which you have to check later, before the final printout is made, but which would disturb your workflow too much at the moment. For example, use a search marker like "###", which is entered into the file at the desired position, where you want to jump to directly. You can add abbreviations to this search marker like "###Sp" = spelling/dictionary, "###Lit" = literature citation or check again bibliographical data, "###Un" = meaning is unknown (look-up the details in a textbook or encyclopedia), "###iS" = introductory or connecting sentence etc.

▶ Note all remaining necessary work steps in your to-do-list immediately, so that you do not forget them.

When proofreading you also check the layout. Figures and tables may not be split apart as long as they fit on one page. The figure subheadings and table headings must appear on the same page as the figure or table. If paragraphs must be split by a page break, no single line should remain on the old or flow to the new page. In addition, a document part heading should not be split from the first paragraph of the appertaining section.

Since the page numbers may change until you entered the very last change, the page numbers are the last thing you enter into the manually created table of contents and the automatically created table of contents must be updated again. This is also true for the list of figures, list of tables etc. Therefore, the master copies of the table of contents and the other lists are the last master copies that you print out. Before you start copying your Technical Report you should check again all cross-references to other document parts and compare all page numbers of the document part headings in all previously printed master copy pages with the entries in the table of contents.

Corrections should always be entered in red, e.g. with a thin felt pen or ball pen. This is important, because you can then see your own corrections far better during entering them into your computer files as compared with corrections marked with pencil or blue or black ball pen. If you have entered a correction, you should move a ruler down across the page until you reach the next correction symbol. You can as well tick or cross out the correction symbols with a pen or pencil having a different color.

Correction symbols are especially important, if the author of the Technical Report enters the corrections on the proof-printout and someone else enters them into the computer files. You can find the standardized correction symbols in ISO 5776. On the internet various people have published their correction symbols. We recommend a meaningful selection of correction symbols with some simplifications, Fig. 3.30.

▶   The symbol to insert a blank line ">" can be completed with "iS>" for an
    introductory/connecting sentence, with "Pb>" for a page break, with "+>" for
    more vertical space and with "–>" for less vertical space.

You can learn and immediately use this simplified system of correction symbols by simply reading the checklist above, since all symbols are logic in themselves and can be deducted again easily. These correction symbols have proven its value in practice very well.

Beside these correction symbols, you can use correction markings, which are to be worked on later. During the phase "proofreading and correcting" you will probably work on the "###" markings in the text with appropriate corrections and delete the markings as well as rework the text style. However, even when you enter the corrections, you may still decide to leave some items unsolved, e.g. because special textbooks or encyclopediae can only be consulted during the next visit to the library. You should structure this phase of inserting the corrections as follows:

– In the printout at the left margin mark all lines containing the "###Sp", "###Lit", "###Un" und "###iS" with a vertical line with a well visible (colored) pencil or text marker. In addition, write appropriate notes what to do into the margin, e.g. in which book you want to read again the marked issue. With these well visible (colored) vertical lines, you can find these editing notes quickly in the end phase, which is characterized by time pressure.
– Cross out all vertical lines in the left margin, as soon as you have entered the correction into your file, if possible with a pencil in a different color like green.
– Still necessary work steps should be ordered by category, copied into a file as to-do-list ordered by category and printed out, resolved items in the list should be crossed out, this should be updated in your file etc.
– Work on the last items of your to-do-list or speak with your customer or supervisor to make sure, that the last changes are not necessary any more.

If possible, let other persons also read your Technical Report. After several weeks or after frequent reading of your own texts you will become "routine-blinded" against your

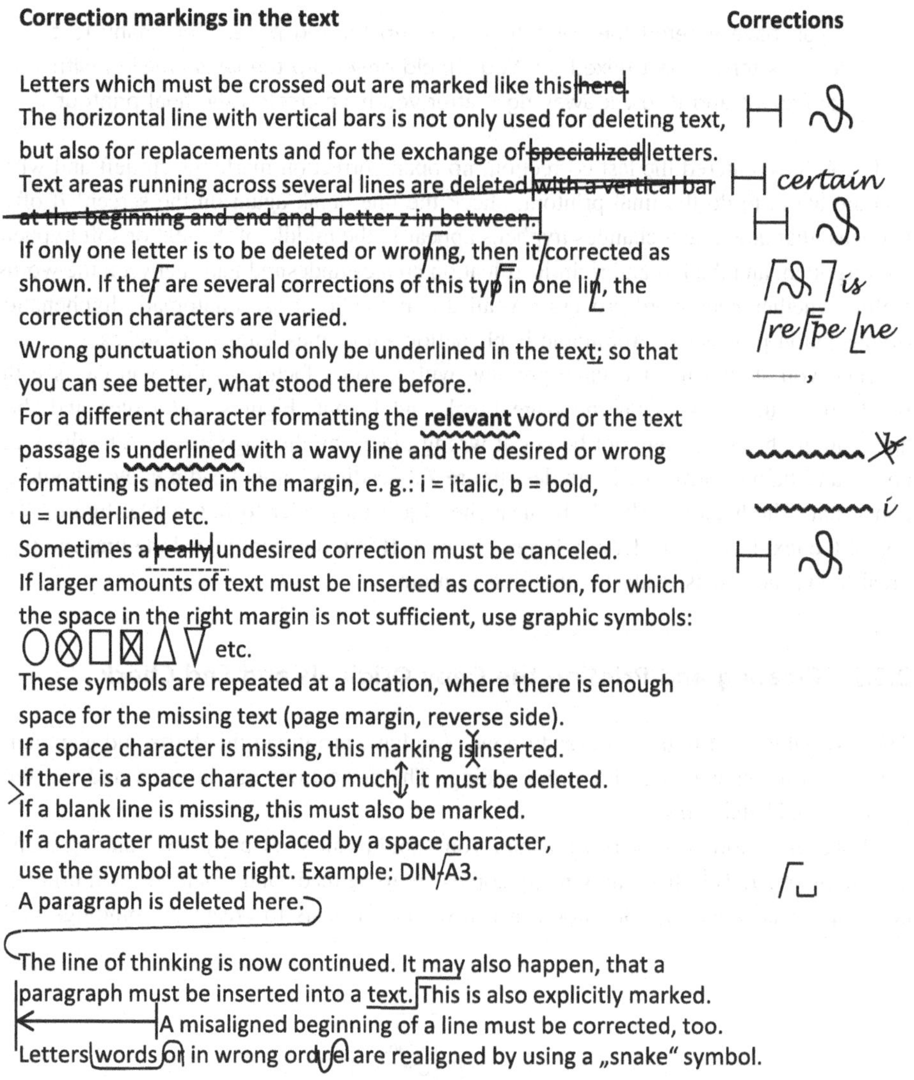

Correction markings in the text   **Corrections**

Letters which must be crossed out are marked like this here.
The horizontal line with vertical bars is not only used for deleting text, $\vdash$ ~~
but also for replacements and for the exchange of specialized letters.
Text areas running across several lines are deleted with a vertical bar $\vdash$ *certain*
at the beginning and end and a letter z in between. $\vdash$ ~~
If only one letter is to be deleted or wrong, then it corrected as
shown. If the are several corrections of this typ in one lip, the $\lceil$~$\rceil$ *is*
correction characters are varied. $\lceil re \lceil pe \lfloor ne$
Wrong punctuation should only be underlined in the text; so that
you can see better, what stood there before. $--$,
For a different character formatting the **relevant** word or the text
passage is underlined with a wavy line and the desired or wrong   ~~~~~ $\cancel{b}$
formatting is noted in the margin, e. g.: i = italic, b = bold,
u = underlined etc.   ~~~~~ *i*
Sometimes a really undesired correction must be canceled. $\vdash$ ~~
If larger amounts of text must be inserted as correction, for which
the space in the right margin is not sufficient, use graphic symbols:
$\bigcirc \otimes \square \boxtimes \triangle \vee$ etc.
These symbols are repeated at a location, where there is enough
space for the missing text (page margin, reverse side).
If a space character is missing, this marking is inserted.
If there is a space character too much, it must be deleted.
If a blank line is missing, this must also be marked.
If a character must be replaced by a space character,
use the symbol at the right. Example: DIN A3. $\lceil_\sqcup$
A paragraph is deleted here.

The line of thinking is now continued. It may also happen, that a
paragraph must be inserted into a text. This is also explicitly marked.
A misaligned beginning of a line must be corrected, too.
Letters words or in wrong order are realigned by using a „snake" symbol.

Fig. 3.30 Simplified system of correction symbols according to ISO 5776

own formulations. The following persons are candidates for proofreading: father/mother, partner, boyfriend/girlfriend, fellow students, supervisor, professor, customer etc. They should have at least one of the two following qualifications:

- expert for the project/specialist and
- expert for spelling or foreign languages/generalist.

Eventually, a different person should do the proofreading for the other area.

▶   If you have entered the corrections, tidy up the old pages. Too many text
    versions tend to get mixed up. You should only keep the latest version with
    corrections and throw it away right after you have created the final printout.

If you have entered the last correction, no open correction markings are left and when
you are about to do the final printout, check the line break again on the screen! It often
happens, that due to text changes hyphens appear in the middle of the text or soft hyphens
are forgotten, and the left and right justification creates undesired gaps between the words.
Define whether your word processor shall do automatic or semi-automatic hyphenation
(hyphenation proposals). Automatic hyphenation sometimes leads to mistakes.

Then you should use the page preview with a zoom factor, so that you can see the
whole page to check again the page break and layout. Figures and tables and their
appertaining headings may not be cut at wrong places by the page breaks. The distances
above and below formulas, examples, figure subheadings, table headings etc. should be
consistent. The headings should be near the object they refer to (e.g. table, figure). The
rest of the text has a considerably larger distance. If you have optimized the line and page
break now, the text is ready for the final printout.

3.8.3   Creating and Printing the Copy Originals and End Check

The text, tables and figures are ready now. You have proofread the drafts and eventually
gave them to other people for proofreading. The corrections are inserted and the final
printout could start now.

However, before you start with that, you should call your copy-shop and announce
your copy order. Tell them, how many copies are to be made and when you will bring the
copy originals. Besides, you should ask how long it lasts to create the bindings (cold

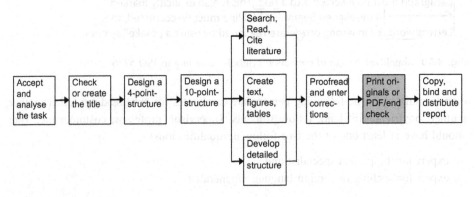

Fig. 3.31 Network plan for the creation of Technical Reports: printing originals or PDF and end
check

adhesive bindings need ≥ 1 day to dry). If you have arranged the copying of larger works or larger print runs with the copy-shop, you can start to create the copy originals. In the network plan, this is the second last phase of creating the Technical Report (Fig. 3.31).

While you can use nearly any paper for drafts, you want to get as high-class copy originals as possible. High-class means, that the copy originals should be rich in contrast. That means at first, that the printing paper should be pure white. That is also valid for figures, which are copied from another source and integrated into your report. In addition, the paper shall not shine through. Therefore, you should use paper with a mass of unit area of 80 g/m^2.

For ink jet printers you should use as smooth paper as possible, eventually even special paper. If you want to integrate a few copied pages into your report or glue-in copied figures, you should use the same paper for these purposes. That means that you should fill the copier with your own paper, if necessary. Do not use pure white paper in your own printer and ecological or recycling paper in the copier. This would result in an avoidable inconsistent impression. Color copies may be on different paper, because you normally cannot influence the paper quality for these copiers.

Manually drawn figures (freehand drawings or drawn with rulers and stencils) appear especially pretty and rich of contrast, if you draw them larger and reduce them with the copier. Small irregularities and blurs disappear. The lines appear much sharper.

The importance of manual drawing declines, but compared with using graphics and CAD programs it still has a certain significance. The phases "creative sketching" and "drawing ready for printing" seamlessly merge when drawing with a drop action pencil and you can rub out errors easily. The copier gets out additional contrast, if you use pure white paper and a high contrast setting. Especially, if you run out of time towards the end of your project, you should seriously check this alternative. In addition, in our experience you can concentrate on drawing with few errors after many hours and late at night better, if you are working manually. With the computer you have quickly pressed a button, you did not want to!

If copied or cut-out figures or tables shall be glued in, always use drawing board and ruler. Clamp the paper and adjust the paper edge. Then lay the figure loose on the paper. Decide, whether the figures shall be arranged left justified, centered, or with a regular offset of a few centimeters from the left edge. Roughly, remember the figure position. Then put as few glue as possible on the rear side of the figure (e.g. only into the corners) and glue the figure in. Now you can use the ruler to check whether the figure is straight. If not, move it carefully. Twist-up adhesive stick (glue stick) is reliable. Compared with all-purpose glue it has the advantage, that you can use a knife to get off the figures again, To adjust them at a new position or to use them in a different context.

The montage glue "Fixogum" has the advantage, that the figures can be removed (stripped off) very easily. You can even get off newspaper paper. However, this glue has the disadvantage, that after four or five years all figures that you have glued in become loose and that the paper gets yellow at the glue spots!

If you use transparency paper, the glue shines through (eventually copy the image or drawing from transparency paper to pure white paper and glue that in).

When copying, the edges of the glued in figures might be visible on the copies. Then you have two options: Change the contrast at the copier or—if the copies may not become brighter—cover the edges on the copy original with correction fluid. You can also make a good, dark copy and cover the undesired edges with correction fluid there. Then this processed copy becomes the new copy original.

For all labels you want to write by hand and colored accentuations of different information (e.g. distinguishing the variants in the morphological box, colored underlining etc.) there are some rules for the selection of drawing pencils, if you do not use the color ink jet printer.

If you draw with felt pens, you can usually see the color also on the rear side. It is better to use text marker, pencil, colored pencil, ball pen and colored india ink. However, ball pens often "spit" (especially before the mina becomes empty) and create thick blurs at the beginning and end of the line, which dry out very slowly and which are therefore smeared very easily.

Refilled text markers also spit very often! India ink pencils create the most accurate results; but since you have to wait for any line that the ink is dried, they take a lot of patience.

Ink ball pens and fineliner pens are easy to handle and have an accurate result. Yet, you should try them out on scratch paper before you start with final writing or drawing.

The colors yellow, light brown, light green, light blue etc. are hard to copy with black-and-white copiers. In the copy, they are either ignored or displayed as very light gray. The colors dark blue, dark green, red etc. can be copied well. They are displayed as dark gray to black.

If only the master copy of the Technical Report shall get colored accentuations, then you should use strong, dark colors and different line styles on your copy originals. On all other copies but the master copy, the different black line styles are clearly distinguishable.

If you want to glue in labels into figures, you should leave at least two spare lines between the different labels on your paper printout. Then you cut out the labels and glue them into the figures. The procedure is the same as described above for gluing in figures. To glue in labels you should use Twist-up adhesive stick (glue stick). Liquid glues flow out at the sides of the narrow paper stripes and excessive glue, which has not been removed in time, glues together the copy original pages or is visible on the copies.

Figures and tables can be larger than DIN A4. This is called "swing-out". Prepare a nearly empty page as copy original for such a picture or table ("carrier page"). On this carrier page, there are only the normal header with a running page number, eventually a footer and the figure subheading or table heading. The figure or table is created in an independent file and printed out or copied (eventually on a color copier) or plotted as often as the number of copies of the final bound Technical Report.

After copying the Technical Report including the carrier page the swing-out is glued onto the carrier page. All left bending edges must be clearly out of the binding range. The right and lower bending edges must have a distance of five (better ten) millimeters from the paper edge, since for some binding types after binding the document is cut with the hydraulic shear at the upper, right and lower edge, see the figures in Sect. 3.8.5.

If you have used a technical part or assembly from a manufacturer brochure in your design, it is better for the understanding of your readers to include the used brochure into an appendix of your Technical Report. If you do that, another step is recommended very much. Help your readers to quickly find the used formula, selected measure etc. by accentuating that in the brochure in the appendix with a text marker.

After all figures and tables are glued in, all manually created labels are written and all correction steps are done the end check can follow:

Are the pages in the right order? Are they upright (page headers at the top)? Pages in landscape format lie correctly, if the upper page margin which is to be bound is on the left side in the pile of sheets and the pages can be read from the right. Have outdated printouts slipped-in? Go through the report checklist, Sect. 3.8.1.

If everything is checked—and also the page numbers in the table of contents have been checked with the copy originals—you can go to the copy-shop.

3.8.4 Exporting the Technical Report to HTML or PDF for Publication

If you want to publish your Technical Report in a data network, i.e. in an intranet or in the internet, there are two popular formats available: HTML and PDF. The PDF format (Portable Document Format) was defined by Adobe.

PDF is a page description language. Every reader sees your Technical Report on the screen in exactly the same way as you have designed it on your computer. The line and page break is kept. That is the main advantage of PDF.

HTML (Hypertext Markup Language) is the page description language used in the internet and in an intranet. Every reader sees your Technical Report as the currently used browser interprets the commands. The line and page break, you have intended, gets lost. That is the main disadvantage of HTML. The main advantage is, that HTML files are smaller than PDF files. Therefore, they can be displayed much faster. We do not want to present an HTML course here and propose the following links instead:

- HTML reference SelfHTML by Stefan Münz: https://wiki.selfhtml.org/wiki/Startseite (in German)
- HTML reference by w3schools.com http://www.w3schools.com/tags/
- online HTML editing: http://htmledit.squarefree.com/
- HTML-Editor Phase 5 is free-of-charge: http://www.phase5.info.
- HTML editor HTML-kit 292 is free-of-charge: http://www.htmlkit.com/
- Coffeecup.com also offers a free version: http://www.coffeecup.com/free-editor/

Even if you work only with office programs, you should know the principle of linking pages in HTML and PDF, because most office programs offer functions to link documents among each other and with external files since quite a while. The following paragraphs describe the most important functions.

Automatically created links (hyperlinks)
The word processors, office programs, and DTP programs—not only creates the features needed for the publication on paper like entries in lists, tables and indexes, labels, cross-references, but in addition it automatically creates the links needed for online publication in the background. You can see this e.g., because all cross-references in the document are clickable and have the paragraph format Hyperlink. In addition, internet addresses are automatically converted to hyperlinks. The prerequisite is that the address either starts with "www." or with "http://".

Manually created links
In your word processor, you need functions to insert a text marker and a hyperlink to link pages. A hyperlink is a connection, a reference or a goto command. You can link each clickable object including images. If you click on the hyperlink, you go to another position within the same file or a program is started, and the file or internet address named in the hyperlink is displayed. The other program can be a presentation graphic program or Real Player, for example. A text marker is the address of a go to command within a file.

Following a link
If you move the mouse cursor over a link in a file, the cursor will change. The cursor shape becomes a hand pointing upwards. In addition, an information text in a yellow box appears. It either tells the link address—then you can click it directly—or it asks you to press the Ctrl key and the left mouse button at the same time to go to the link address (depending on the program version). This holds true for all Microsoft office and open office programs and many other programs which you can use to display and edit text files, slide presentations and calculation sheets. By the way, the cursor shape changes in the same way, if you move it over a link in an HTML or PDF file.

Export from your word processor to HTML and PDF
The references, which you have manually defined in a text file and the automatically created references remain 1:1 unchanged, if you save the file in HTML format or in PDF format. When exporting to HTML, each text file results in one HTML file. You have to manually insert links between the files or manually create an HTML ToC file.

When exporting to PDF, please keep in mind the following: Using the conversion options it is possible to set the access permissions to the PDF file, e.g. you can allow reading only and forbid printing. If the created PDF file shall contain your document part headings as a tree structure to navigate at the left side of the file, you have to specify in the conversion settings, that bookmarks shall be created and which headings and other paragraph format templates shall be used for the creation of bookmarks.

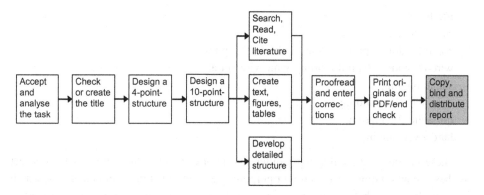

Fig. 3.32 Network plan for creating Technical Reports: copy, bind and distribute the report

3.8.5   Copying, Binding or Stapling the Technical Report and Distribution

Copying, binding or stapling and distribution of the Technical Report is the last step in the network plan for the creation of Technical Reports, Fig. 3.32.

If you want to distribute or publish your Technical Report online, you have to save the final version of your data again in PDF (or HTML) format and send the files to your webmaster.

If you want to distribute or publish your Technical Report in hardcopy form, it has to be copied or printed first. In most cases, it will be copied, because the print run is small. All copies shall be made on the same copying paper. This is also true, if several authors work together to create the Technical Report. One single exception is admitted: There are color copies in your Technical Report. In this case, there is usually different paper required because of the different copying process for color copies and therefore it is admitted.

After copying, the report is bound. This makes a readable document of it, which—depending on the number of pages and binding type—can have the character of a script, booklet or book.

In the copy-shop, you should stay a while in case there are open questions. There might be questions like: Shall we align the cover sheet a little different or enlarge it a little? Which plastic spiral (comb) or binding wire or fabric tape is desired? Which cardboard with which surface structure and color shall we use? etc.

The following stapling and binding types are available:

- staple/paper-clip
- saddle-stitching (=binding type of journals)
- plastic folder
- filing fastener (=plastic stripes with holes and metal tongues)
- spring binder
- spring strip

- file binder
- ring binder
- comb binding (spiral binding with plastic spirals)
- wire-O binding (spiral binding with wire spirals)
- staple binding
- cold adhesive binding
- hot adhesive binding
- hardcover binding

These stapling and binding types differ a lot in price and handling properties. To select the best suited binding for the current purpose, we want to list those points at first, which have to be done before the binding, and then we shortly describe the properties (advantages/disadvantages) of the various bindings. For determining the binding type, you should speak with your customer or supervisor.

To be done before the binding
If you want to bind swing-outs (=figures, tables or lists, which are wider or larger than DIN A4): are the bends on the right and at the bottom far enough away from the paper edges (≥ 5 mm), so that they do not fall into pieces when the report is sheared? When the report is shaken in the copy-shop, they have to make sure that the pages with the swing-outs move into the binding area correctly! This is especially important for adhesive bindings. Problems can be avoided, if you let them bind a DIN A4 carrier page into the report and glue the swing-out(s) onto the carrier page(s) after the binding.

If you have drawings, tables, figures or lists, which are larger than DIN A4, you have to decide and keep in mind the following points:
    Swing-outs to the bottom should be folded and glued as shown in Fig. 3.33.
    Shall copies of drawings be bound into the Technical Report?

- Drawings in DIN A2 and DIN A3 can be reduced to DIN A4 with the copier without problems and normally bound.
- Larger drawings should be reduced with a copier by maximum two DIN steps (e.g. DIN A0 to DIN A2). Then fold the copies to the normal paper size DIN A4 according to DIN 824, see image below.
- The unfolded original drawings are always handed in in a folder or roll.

If there are DIN A3 drawings in landscape format or swing-outs to the right, you can decide, whether you want to bind them into the report or combine them eventually together with drawings of other size(s) in a file folder. If the drawings shall be bound

Fig. 3.33 Folding of a
swing-out to the bottom

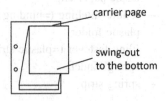

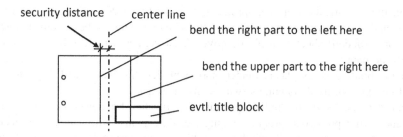

Fig. 3.34 Folding a DIN A3 drawing or a swing-out to the right to DIN A4

together with the report, fold them as shown in Fig. 3.34. The security distance assures, that the pages of the Technical Report can easily be turned. If you bind this drawing as swing-out into the report, the security distance helps, that the swing-out is not destroyed during shearing,

If you want to fold larger drawings according to DIN 824, you can only present them in a file binder. They cannot be bound by staple binding, cold or hot adhesive binding, or hardcover binding, because the drawings folded according to DIN 824 would be cut into many pieces, when the volume is cut for finishing the binding.

If the drawings shall not be bound into the report, but presented in a file binder, fold them according to DIN 824, Form A as shown in Fig. 3.35.

The folded drawing (or plan or table or figure) must be unfoldable and refoldable while it is still bound in the file binder. Plastic adhesive reinforcing rings are recommended. You will find more details regarding folding the paper formats DIN A0, A1, A2, and A3 in DIN 824, esp. exact dimensions.

If you want to glue in photos, copied images or images from other printed materials (newspapers, journals, brochures etc.), please do that before binding!

If you want to use sheet protectors, make sure to use a type that has three closed sides with the open side at the top.

Sheet protectors can only be used together with the binding types plastic folder, filing fastener, file binder and ring binder. Sheet protectors cannot be bound into staple binding, the adhesive bindings and hardcover binding, because they would be destroyed during shearing.

Did you sign yourself the declaration in lieu of an oath after the copying with a document-proof pencil (e.g. ballpen or pen)? All copies must be manually signed.

The binding types staple/paper-clip, filing fastener and spring strip do not provide a container for the sheets of paper. If you have decided to use staple or paper-clip, you can

Fig. 3.35 Folding a DIN A1
drawing to DIN A4

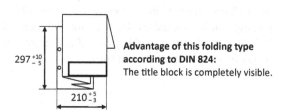

297^{+10}_{-5}

210^{+5}_{-3}

Advantage of this folding type
according to DIN 824:
The title block is completely visible.

hand in the documents in a sheet-protector (one or two sides open). In case of filing fastener and spring strip the first and last "sheet" can eventually be a (semi-)transparent plastic or cardboard sheet or a plastic cover. That creates a more tidy impression and protects the paper.

For the binding types plastic binder, spring binder, file binder and ring binder you should not hand in used containers (with signs of wear).

Often it is useful to bind bulky appendix material like brochures, corporate publications, measuring protocols, program listings, electric plans, wiring diagrams, technical drawings etc. in a separate file binder or to bind them separately. This separate binder belongs to the report. The documents are listed in the list of references. These documents are also added to the table of contents of the Technical Report with the note "(in separate binder)". If you would have to list many documents with that note, it is better to split the table of contents into one for volume 1 and one for volume 2. Both tables of contents are then included into both volumes of the Technical Report.

Now the properties of the various binding types are described.

Staple/paper-clip
Advantages: Extremely cheap, after removing the staple or paper-clip the documents can be punched and integrated into the filing system, copying with paper feed is easily possible.

Disadvantages: Only suited for thin documents, to read the inner pages you have to bend the corners. If there are many documents with staples or paper-clips in a folder, the upper left corner becomes very thick.

Saddle-stitching (two staples in the back of a brochure)
Advantages: This binding is used for journals, brochures and booklets of any type, is extremely cheap and requires, that the Technical Report has been reduced during copying (from DIN A4 to DIN A5) and copied double-sided.

Disadvantages: This binding is created with a stapler with longer arm, then the complete report is folded once in the middle. It is possible to process max. 20 sheets (80 pages) with this binding type. If you want to copy the bound report later, you have to manually turn the pages and put the report onto the copier, if the binding shall not be opened.

Plastic folder
Advantages: This binding is very cheap and easy to be done. Sheet protectors can easily be integrated. With this binding type you can bind up to about 100 sheets; if there are more sheets, the report becomes more and more difficult to handle.

Disadvantages: To read the inner pages, you have to bend the pages to the left, because otherwise the report would not stay opened when lying on a table. The metal tongues are often sharp-edged, if the report is read more often, the holes become frayed. Since the binding can be opened easily, copying with paper feed is easily possible, as long as the holes are not too frayed. Therefore this binding type is only suited for your own documents or for unimportant purposes.

Filing fastener

Advantages: Filing fasteners are much cheaper than plastic binders. The handling of the bound documents is comparable with plastic binders.

Disadvantages: The filing fastener does not provide a container for the sheets, it is only possible to file the filing fastener with its cardboard or plastic stripe, onto which the bound sheets are fixed, into a file binder. Filing fasteners are not suited to present your Technical Report in an optically attractive way. Therefore this binding type is only suited for your own documents or for unimportant purposes.

Spring binder

Advantages: Spring binders are available in different variations. The sheets are not punched, but put into a protecting plastic cover and clamped together with a plastic bar with a groove (=spring strip). The spring strips are available in different thicknesses. If the sheets shall not be punched, but still presented in an optically attractive form, this binding type is suited well (like sheet protectors). Further copies can be made easily with paper feed.

Disadvantages: The number of sheets is limited to about 30 to 40 sheets. sheet protectors are hard to clamp, because they are too smooth. The bound report does not stay opened. It shuts itself again.

Spring strip

Advantages: The spring strip is similar to the spring binder, but it is much cheaper, because it does not have a plastic cover.

Disadvantages: The spring strip does not provide a container for the sheets. It only provides a losable connection of the sheets.

File binder

Advantages: You can turn the pages well, the pages remain opened, sheet protectors can be integrated easily, copies can easily be made with paper feed, it is possible to bind comparatively many sheets (in 8 cm file binders you can bind up to 500 sheets without staples and paper-clips).

Disadvantages: File binders usually need much space on the shelves and are rather bulky in opened state. Besides, if you often turn the pages, the holes may become frayed. The only countermeasure is to stick on plastic adhesive reinforcing rings.

File binders are not very attractive for presentation purposes. Therefore select a different binding type for your Technical Report and use a file binder only for an appendix (or several appendices) with brochures, manufacturer catalogues, corporate publications, measuring protocols, program listings, drawings, plots or copies of your drawings which are larger than DIN A3 and folded to DIN A4 according to DIN 824.

Originals on transparency paper are never folded! They are stored and transported in drawing folders or rolls. During meetings the originals can be fixed at a clamping or magnetic bar at the wall in the meeting room (but even there plots or copies are better/safer).

Ring binder

Advantages: Ring binders have for metal rings in Europe and three metal rings in the US which can be opened to insert or take out punched sheets of paper. They provide an optically attractive presentation of smaller Technical Reports. Commonly sheet protectors are used and the sheets are inserted into the protectors (three sides closed, top side open), so that you can read the pages like in a book. Then the pages with odd page numbers should appear on right sides and the pages with even page numbers on left sides (like in a book).

Disadvantages: Ring binders are usually more expensive than file binders. Inserting the sheets into the sheet protectors takes effort!

Comb binding (spiral binding with plastic spiral)

Advantages: This binding type is relatively compact and cheap. You can put a sticky label with the name of the report (and eventually author) onto the plastic spiral/comb and the sheets stay opened when the report is lying on a table, you can turn the pages of the bound report around nearly completely. With appropriate spirals/combs this binding type can bind up to about 500 sheets.

Disadvantages: The sheets must be aligned carefully before they are punched and during punching, they must be laid against the stop bar of the machine with care, so that the holes are not too near to the paper edge and tear off. Caution! Do not punch too many sheets at the same time. If you need too much force for punching, the machine might be damaged!

Commonly a transparent cover (thicker than overhead slides) is used on top of the title leaf. Caution! This sheet must be punched separately in any case. Otherwise, the transparent cover and the paper sheets move against each other and the punching is irreparably wrong.

In the copy-shop, the sheets are shaken on a machine. There it is also possible to shear the bound Technical Report at the right edge, because the sheets are not lying exactly on top of each other, but they adopt to the curvature of the spiral/comb.

If the spiral/comb is not large enough, the sheets are bent near the rectangular binding holes and then you cannot turn the pages very easily any more. However, if the spiral/comb is much thicker than the Technical Report, then the report does not stand well on the shelves.

The plastic spirals/combs are often sharp-edged at the corners. You should round off sharp corners with a pair of scissors before using the comb. If you do not have a machine for creating the binding (i.e. punching and opening and closing the combs), you can only make copies by manually turning the pages and putting the report onto the copier.

Wire-O binding (spiral binding with wire spirals)

Advantages: The wire-O binding is even more compact than the comb binding and about equal in price, but the sheets do not bend so easily when turning the pages, because the binding holes are round. With this binding type, books can be turned around completely. Therefore, this binding type is well suited for manuals. With this binding type, you can bind up to about 350 sheets.

Disadvantages: Putting a sticky label with the name of the report (and eventually the author) is not possible. The information regarding a transparent plastic cover sheet given under comb binding is valid here as well. For wire-O binding it is also true, that you can hurt yourself (here at the wire) at the upper and lower edge of the report. Chamfer sharp edges with a file. Copies can only be made by manually turning the pages and putting the report onto the copier.

Staple binding
Advantages: This binding type is a little cheaper than the spiral bindings. The stack of sheets is shaken with a machine. Then five staples are twinged through the stack and bent on the other side. Then a fabric tape is attached to the binding, which covers the staples, a small stripe on the front and rear side and the spine. Then the book is sheared with the hydraulic shear at the three open edges. This binding type is extremely stable. The impression is created, that you are holding a real book in your hands. Due to the stability and the attractive appearance, this binding type can be recommended for Technical Reports. For theses and smaller reports during a study course, the staple binding is the most common binding type. With this binding type, you can bind up to about 200 sheets.

Disadvantages: The sheets/the book does not stay opened when lying on a table. Therefore in most cases only the front sides of the pages are printed. Copies can only be made by manually turning the pages and putting the report onto the copier.

Cold adhesive binding
Advantages: This binding type is a little cheaper than the spiral bindings. The stack of sheets is shaken with a machine. Then the spine is covered with cold glue, which needs to dry for one to two days. Then a fabric tape is attached to the binding, which covers a small stripe on the front and rear side and the spine. Alternatively, a printed cardboard envelope can be attached, which covers the front side, the spine and the rear side. In this case, you can easily write onto the spine whatever you like. Then the book is sheared with the hydraulic shear at the three open edges. The impression is created, that you are holding a real book in your hands. With this binding type, you can bind nearly any number of sheets. Due to the possibility, that you can also bind thick books and use a cardboard envelope as well as due to the attractive appearance this binding type can be recommended for Technical Reports.

Disadvantages: Depending on the stiffness of the glue, the sheets do not stay opened when the report is lying on a table. Copies can only be made by manually turning the pages and putting the report onto the copier. This binding type is not very stable. It happens often, that a few pages fall out of the glue, especially when the report is copied.

Hot adhesive binding
Advantages: For this binding type, the sheets are put into a folder/an envelope, which has a hard stripe of glue in the spine. The folder/envelope is put into a machine which heats up the glue so that it is pasty. Then the folder/envelope with the sheets is taken out, and the spine is stuck onto a hard underground, so that the paper sinks even deeper into the glue. Now the glue needs to cool for a few minutes to become hard again. This binding type is very attractive for smaller reports, e.g. seminar documentations.

Disadvantages: See disadvantages of cold adhesive binding.

Hardcover binding
Advantages: This binding type is the "most valuable" binding type. It is stable, the sheets of the book remain opened, when the book is lying on a table and they do not fall out of the book. The pages are sheared and the cover can be printed or embossed as you like.

Disadvantages: This binding type is very expensive. Therefore, it is usually out of choice for Technical Reports. The only exception might be one copy of a thesis that is produced e.g. with black cover and golden letters. This copy is well suited to be shown later in an application interview.

In the next chapter, we have collected hints about a useful behavior for working on your project and writing the Technical Report. These hints are grouped into the areas working together with the supervisor or customer, working together in a team, working in the library, paperwork, file administration and personal working methodology.

3.9   References in This Chapter

Since this is the main chapter of this book, a separate listing of the references in this chapter is not useful. Please refer directly to the reference part of this book.

Useful Behavior for Working on Your Project and Writing the Technical Report

4

▶ In this chapter, we will introduce some useful information and procedures to improve your working effectivity. The following sections contain hints for working together with your supervisor or customer, working together in a team of authors or a project team, working in a library, organizing your paper documentation and the computer files, creating back-ups of your data, as well as hints for your personal working methodology while writing your Technical Report.

By writing your Technical Report, you provide a job performance in professional form and methodology that shall bring you success and reputation of your abilities, if possible in combination with going the next step in your professional and scientific career. To achieve this, it is inevitable to make clear which practical and knowledge-based performance you have provided yourself, e.g. executing and evaluating your own experiments, statistical inquiries or providing a summarizing and comparing overview of correctly cited facts, figures and findings published by other authors, which you newly combined to shape new own and general findings.

In this context, we would like to recommend reading Chap. B. It does not only contain explanations of technical terms used in this book, but also many definitions of technical terms in printing technology, which may come across when you are working together with copy-shops, computer stores, printers, as well as journal and book publishers.

In addition, here are some links to the internet, which are probably useful for you when you are writing Technical Reports:

- http://dict.leo.org and www.dict.cc: dictionaries in various language directions
- www.systranbox.com/systran/box: translation of short texts in various directions
- http://www.google.de/language_tools?hl=en: translation of short texts in various directions.

© Springer-Verlag GmbH Germany, part of Springer Nature 2019   159
H. Hering, *How to Write Technical Reports*,
https://doi.org/10.1007/978-3-662-58107-0_4

4.1   Working Together with the Supervisor or Customer

You should always try to prepare and to plan the meetings with your supervisor as well as you can. This includes the following aspects:

Always keep the text and the structure at hand
Always take with you the current version of the text and the current structure of your Technical Report when you go to a meeting with your supervisor. This helps him/her to improve the supervision and support very much, because he/she can then overview the current stage of your Technical Report or project better and faster.

Note questions
Before your meeting, you should write down questions and tick off those questions, which have already been treated, during your talk. This helps to avoid that you forget something.

Prepare a checklist
If you want to take with you many documents and parts for the meeting with your supervisor, a checklist might be helpful, where you list all necessary documents, parts and pathnames to files, which you want to discuss with your supervisor on-screen. This can include the following documents, eventually as files:

- written task,
- master copies from technical documents,
- technical drawings, on which you build up your work,
- manufacturer brochures,
- current version of drafts (on cardboard or as CAD drawing),
- literature sources.

As a precaution, you should also take documents with you to the meeting or provide them for correction, which you think you will probably not need. Please do not present your Technical Report in sheet protectors or provide water-soluble transparency pens.

Provide projectors and equipment for the presentation and try them out
When collecting items for the checklist, you should also regard required projectors and boards like: beamer for the presentation of files, overhead projector, slide projector, flipchart, poster hanging clips, or magnetic bar to fix plans or drawings etc. You should prepare the meeting in such a way, that all required equipment is available in time. If you want to project pictures make sure, that the meeting room can be shaded. Try out the projection conditions prior to the meeting. Can everybody see the images well?

Comprehensive presentation or simple meeting?
You should also think about whether a more comprehensive presentation of the results developed so far is necessary. If yes, you should plan this part according to the information provided in Chap. 5.

Negotiation tactics
During discussions with your supervisor, you should always use the yes-but-approach. That means, even, if you have a different opinion, you should agree at first, so that there

does not occur any sharpness or denial in the talk. After two or three additional sentences you can carefully change to "no" and wait for the reaction of your supervisor.

Compromises and quality defects

Do not make compromises. Even, if your supervisor does not look into every detail at once, you should stick to this rule: Items, where you are doubtful yourself, are nearly always a source for criticism of the supervisor or customer. This is an often proven fact. Therefore, you should not leave formulations you are in doubt about, just because of time-pressure or a research is only possible during your next visit to the library.

If you are uncertain about a statement, think about the basic rule: "All details and statements which are not written down, cannot be wrong." You should either leave that detail or statement out or research thoroughly, to smooth the uncertainties.

Note hints and open action items

During your talk write down any hints from your supervisor and action items still to do. Think about whether these hints refer to a meaningful specialization of your project, additional literature resources, or new sources of information and take appropriate short notes.

Plan the deadline

Towards the end of your project, you should try to only work on hints of your supervisor, which do not cause too much work. If there are hints causing a high workload, you should point out the consequences in time and decide for yourself, how much additional work you are willing to do. However, you should negotiate that as open and early as possible with your supervisor.

4.2   Working Together in a Team

Here we want to discuss problems, which only occur, if several authors write different parts of the Technical Report. At the beginning of your project, you can discuss the following central questions: Under which conditions are we working? What do we want to achieve? Who does what? Who has taken over responsibility for what? What are the special abilities of the group members? Who needs to learn what? Who needs to obtain what? What are our working procedures? What, how and with whom do we communicate inwards and outwards?

Corporate rules and standards

Do you know all corporate rules and standards of your university or company or do you still need to find out details regarding page layout, template files etc.? If yes, please care for the details in time! If there are computations, you should define in advance and eventually discuss with your supervisor, how many decimal places shall be used in formulas.

Report checklist and style guide
For teamwork the usage of the report checklist and style guide are especially important.
Otherwise one group member writes "figure 3 to 8" and another one "figure 3 thru 8" and
these inconsistencies do at least bother your readers.

Structure
Agree on a common structure before you start to write the text. The structure should be
finer than a 10-point-structure, if possible. Changes to the structure should be communicated to the other group members as soon as possible.

Common technical base
If you are all using the same version of word processor program and the same formatting
templates and the same fonts, your systems are already rather compatible. In addition, you
should use the same printer driver for the (final) line and page break. Please also use
common fonts so that special characters and symbols will not look completely different
than planned on the computer on which the final version of the report will be printed. This
holds also true for slide presentations, especially for the bullet symbols.

Consistent layout
The page numbering often does not fit together, because all group members use their own
page numbering. Therefore, there are page numbers, which occur more than once, or gaps
in the page numbering and erroneous cross-references. This needs to be adopted before the
final creation of table of contents, list of figures and list of tables. In any case, we
recommend that the group member with the best computer literacy combines the text parts
written by the different group members and then prints out the report, because he/she can
then still correct mistakes of the others.

4.3   Advice for Working in the Library

In Sect. 3.5 we have already described, that working with literature is important to prove
or disprove your own position, to get new ideas and to reflect the current state-of-the-art.
Yet, working with literature is very time-consuming. Therefore, you should try to minimize the time it needs by intelligent planning and organization. Here are some hints how
to do that.

Guided tour or introduction to library usage
If you are not so versed with your local library, you should take part in a guided tour.
There you learn, how the various types of literature are searched (online catalogues,
microfiche catalogues, card indexes for older publications etc.), where the literature is
located according to the location label and signature (on reserve shelves or in the depot,
which corridor for which field of knowledge) and whether, when and how the literature
can be borrowed (computer-aided loan, depot order, distance loan, short loan etc.).

Small change, copy card, paper clips, note taking
Recent journals may not be borrowed in most cases. Articles from them must be copied in
the library. In addition, the reference copy of desired books may not be borrowed in most

cases. To divide several stacks of copies, you will permanently need paper clips. It is best, if you put some paper clips into a little plastic box or bag, e.g. an empty film box. In addition, you should take with you at least two well-functioning ball pens and the notes collected for the next visit to the library. At home, you should preferably put these notes into a firm DIN C4 envelope, labeled with "library".

Order of work steps in the library
If you have arrived at the library, you should perform all work steps in the following order:

- switch your mobile phone from sound to vibration alarm
- internal depot orders (last about 1 h: you can take the books with you!)
- external depot orders (last about 1 day: you cannot take the books with you!)
- search for literature on reserve shelves (the books are on the shelves)
- eventually search for literature in the dissertation archive, search for literature on microfiche, (let someone) make enlarged copies from microfiche etc.
- visit to a library presenting standards and patents to the public (the library employees will help you to find technical rules and guidelines, standards and eventually patents)
- reading literature, making copies
- working yourself through the notes in your envelope "library" and further working on the text of your Technical Report to bridge waiting times.

Note bibliographical data
When you make photocopies in the library, a very important working principle is, that you should note at once all bibliographical data and the data where to find the book in the library on the copied sheets. At least you should note author and year onto the front side of the first sheet in the stack, if this is not already printed there, and write the rest of the bibliographical data eventually onto the rear side of the first sheet of the stack. Then you should connect the stack with a paper clip, mark the author's name on the front side of the first page with a pen in a different color (because the names are printed at different positions in different publications) and then make photocopies from the next literature source or article from a journal. If you proceed like that, you will have all required information at hand, when you write the list of references and you can alphabetically sort the copies and put them into file binders quite easily.

Check copies for completeness
You should not give back the literature immediately after making the copies in the library. At first, you should check the copy stacks, whether there are pages missing. Only if you are sure, that all pages you wanted to copy are completely there in the right orientation and order and with the complete bibliographical data and information where to find the literature in the library, you can give back the literature with ease. Check again, whether there are still open notes in your envelope "library", which you want to deal with.

At home you can put the copied literature either onto the book pile of the appertaining chapter or subchapter or you can punch the copies and add them to your report binder at the right position, see Sect. 4.4.

With the approach described here, you will save a lot of time. Besides, with this working procedure you will lay the foundation for being able to create a correct and complete list of references without having to revisit the library, just because there are some bibliographical data missing. In the time saved by this approach, you can perform the other necessary work steps more carefully and do the proofreading more intensively.

4.4 Organizing Your Paperwork

In this section you will get to know a working method how to sort and store different cover sheet/title leaf versions, drafts for the structure, literature sources, brochures, encyclopediae, notes, text drafts and other materials assigned to your Technical Report. However, every project is built up a little different and therefore needs appropriately adopted techniques to perform the project work and to organize the paperwork.

Report binder
Install a report binder, in which you will collect everything that belongs to the Technical Report. Label it at the back of the file binder with the working title of the report.

Subdivide the report binder with dividers made of firm cardboard, which exceed the paper sheets at the right margin and which are labeled with soft pencil. This has advantages, as long as you are not yet very deeply involved in your project. At this early time you do not yet know all details, how the information should be ordered and structured best. Therefore it may occur that you want to label the dividers differently in the future. This can be done easily, if you use a soft pencil and a rubber.

The report binder may for example get the following structure:

- cover sheet/title leaf and structure versions, the latest version is at the top
- other parts of the front matter, e.g. list of figures and list of tables, eventually with manual additions since the last printout
- printout of all chapters of the Technical Report including the appendices
- material collection (text drafts, notes and photocopies) ordered by chapter
- notes and other material which cannot (yet) be assigned to a chapter
- cover sheet/title leaf versions, the latest version is at the top
- style guide
- report checklist
- to-do-list with action items (buy or do something, list of open corrections).

▶   A copy of the current structure is also always visible lying on or hanging next to your desk.

Book pile
All documents in the material collection, which cannot be punched easily, but which shall be cited (text books, reference books, figure-tables, manufacturer catalogues etc.), are stored on a separate book pile for each chapter, subchapter or section, depending on the

amount of documents. In case of single copied pages and thin brochures, which will remain in your personal archive, you have to decide, whether these documents shall be stored on the book pile or in the report binder. The individual book piles get a cover sheet with document part number and title and a divider (sheet of paper) to distinguish literature already cited from literature which still needs to be processed. Literature, which you do not need any more, is at the bottom.

It happens sometimes, that you (or someone else) wrote notes into documents from the material collection, and you decide later, that these pages shall be included as copy into the text of the report or as brochure into the appendix. It is much better, if you do not write notes directly into documents, which you have not, written yourself, but if you use note sheets or sticky-notes.

Version control

If you have many files, you can label your printed texts and graphics with path and file name, version and date. This is also true for printed cover sheet/title leaf and structure drafts. If you want to glue figures or tables onto your printouts, an appropriate white space remains unprinted. The figures or tables are inserted into the report folder in front of the relevant page of the draft text, e.g. in a sheet protector, but not yet glued onto the text page, because until the final printout there still may be several draft versions and the gluing would be a waste of time.

Proofing

The text draft should be proofread on paper, because typing errors will then be found much better. Scripture can be read more clearly on paper than on the screen. Moreover, text paragraphs can be re-ordered much better on paper than on screen, if there is an illogical sequence of thoughts. Corrections should best be marked in red—even in your own drafts. Corrections and editing remarks in this color are much better readable, when you enter them into the computer, as compared with corrections marked with pencil or blue or black ball pen.

To-do-list

Analogously to the project notebook (jotter) described in Sect. 2.5 for the practical work steps in your project, you should create a to-do-list for writing your Technical Report and note all items like jobs that still need to be done, literature that needs to be bought or borrowed in the library etc., as soon as the items occur. This holds also true for corrections, which you have recognized, but which you cannot or do not want to do at the moment. Prior to the final printout, you should look again through the to-do-list and process all left open items.

4.5   Organizing Your File Structure and Back-up Copies

If you create texts for your report, the first thing that must be defined is, how you want to store your data on your hard disk. It has proven to be practical that you create a subdirectory somewhere in your file tree, having the same name as the working title of your report eventually in abbreviated form. In this subdirectory, you can store the files representing your chapters, subchapters or sections.

File names
Please select meaningful filenames, you can remember easily in the future. Independent of the operating system of your computer the following basic rule has proven its worth.

▶   In file names for text and image files you should not use space characters
        and umlauts. You should only use small letters, numbers and the special
        characters—and _!

This is especially important, if you are working in a heterogenous network with different operating systems (e.g. Windows, Mac OS and UNIX or Linux resp.), since in such complex networks file names are sometimes modified when copying the files so that capital letters become small letters (first letter as capital letter, the rest are small letters), space characters are replaced by special characters (e.g. %20 under Windows) etc. If you search your files from a computer with a different operating system, the searched file eventually cannot be found immediately. Links referring to internet or intranet pages do not work etc.

Structure file
Since the structure is growing while you are writing the text, i.e. it is refined all the time, you should note the following information in the structure in one line of text at the bottom of the file

– the path and file name of the structure file and
– the exact date when the structure was last modified (the version date)

This line of text is then always printed together with the structure.

In addition, the file names of the structure files should get the exact date or month and year when the structure was created. To see the structure files in the correct order, you should write the year at first, then the month and then the day.

Different versions of the structure file

```
struc-casting-system-2009-07-12.doc
struc-casting-system-2009-09-27.doc
struc-casting-system-2009-09-30.doc
```

Keep all older versions of your structure as files on your hard disk and as paper printout, so that you can go back to a previous structure version. This may be necessary, if you cannot get enough literature in the remaining time for the focal point of your project, which has been defined in the last structure version, or if experiments cannot be performed in time, because there are supply difficulties or defects in your equipment, or the supervisor prefers a previous version of the structure.

Order of the files

To see your text files in the correct order, you should specify the chapter and eventually subchapter number in the beginning of all text file names, eventually with with a leading zero for one-digit chapter numbers and then the title of the chapter or subchapter.

By applying this naming convention the files are listed in the same order as in the structure and you keep a good overview. Example (please refer to the table of contents of this book):

Files appear in the computer in the same order as in the structure
01-intro.doc   (chapter 1)
02-planning.doc   (chapter 2)
31-creation.doc   (subchapters 3.1 to 3.3)
34-creation.doc   (subchapter 3.4)
35-creation.doc   (subchapters 3.5 to 3.9)

Graphics files

If you use image files for your Technical Report, you should not only store these files embedded in your text or slide presentation files, but also in their original format, in the original size and original resolution and, if applicable, in a vector format.

You will often need the image files for your Technical Report also in a different format like gif or jpg or in a different size (in pixels) for publishing them in data networks. A different resolution may occur, if you use your images with 300 dpi for printing them in a journal, with 150 dpi for printing and binding the Technical Report and with 72 or 75 dpi (= screen resolution) in a slide presentation to get a reduced file size.

▶   The graphic files should always be archived and passed on to co-workers and publishers in all used file formats, image sizes and resolutions. This is the best way to guarantee that you (and others) can still edit the graphic files later. In case of vector files, this is even the only possible way to be able to properly edit the files later.

Back-ups

Create back up copies of all your data quite frequently! So, you save the results of your own intellectual work. In addition to the files on your hard disk, you should at least create two or three copies of all files belonging to your project on different storage media like USB flash drive, CD/DVD, exchangeable disk, tape, etc.

Update the back-up copies of your files quite frequently (once or twice a day). To pack your data and to compress/uncompress your files under Windows you can use e.g. WinZip. Mark the files to be stored, and click them with the right mouse button, then select "Send to zip-compressed folder" in the context menu and specify a file name.

In addition, keep the last printout of your files in the report binder. This is not exaggerated carefulness, but a working rule derived from bitter experience.

Management of internet links

If you are surfing much in the internet, store your most visited pages as favorites or bookmarks and delete the temporary internet files regularly, so that your computer does not collect several MB of data and becomes very slow, before you do something against it. In the Internet Explorer, there are several interesting commands to do that (depending on the version).

▶ After each larger internet search, you should execute the following commands:

    – Export your favorites or bookmarks in your project directory as HTML file. Save this file regularly together with the other files in your project directory.
    – Delete the browser sequence or browser history.

4.6   Personal Working Methodology

In the literature there are many tips about personal working methodology. For creating Technical Reports these tips are also valid in principle. However, since we deal with "technology" there are a few special aspects. The most important rule is: Plan enough buffer times!

Time planning

The effort needed for actually finishing the Technical Report is regularly underestimated (even if experienced authors make a careful, conservative estimation). Therefore, estimate the required time realistically and multiply the result by two. So you will have a sufficient buffer time.

Then you should create a time plan, e.g. in a Gantt chart, and regularly check whether you are still within the planned time. You can get various Excel based templates for Gannt charts in the internet. You should apply strict self-discipline, even if your project and the work on your Technical Report do not run smooth for a while, but you should also plan small breaks and rewards for reached milestones. Do not lose your working direction in exaggerated perfectionism. It is better to plan an editorial deadline for yourself and keep it.

You should begin with writing the text no later than after 1/3 of your practical project work. Then you can finish the project early enough, to have time for preparing the written examinations at the end of the term or semester.

Buy or get required material and template files in time

If you want to finish the Technical Report during the semester break, you should check whether the paper shop is closed and buy paper, card board, printing ink, toner, forms, transparency paper etc. well before the date of submission or before holidays.

In addition, you should copy files and forms offered by your professor (like part list forms, example entries to part lists etc.) early enough, so that you do not get time problems during the Christmas break or summer break. You should apply this approach for all default documents and default procedures provided by your supervisor, i.e. also for example computations, special diagrams, work sheets etc.

Export and comment internet favorites or bookmarks
If you have to search much in the internet for your project, save interesting addresses as favorites. You should think of managing your favorites in subfolders sorted by topic. It is useful to export your favorites from time to time (in Internet Explorer: File—Import and Export). You will get an HTML file with clickable links and can enter comments and intermediate headings providing a structure to all the links. You can also use this file for team meetings or the next visit to your supervisor or customer.

Start working coarse, refine later
During the initial stage, you should not invest too much effort for the optical appearance of your Technical Report. Reworking the line and page break should best be done only once but thoroughly during the final stage! Formulating the text and selection or creation of figures and tables should always run parallel with writing text. Unfortunately, drawing images lasts quite long! If you postpone this to the end, time problems are inevitable!

Handwritten notes and sketches are absolutely sufficient in the initial stage, to overcome the first writing barriers and to get a feeling for which information should be presented in which document part. If the Technical Report is then written in a first version on the PC, you can mark text parts, which are not-yet-ready.

Not-yet-ready text parts may occur, if you want to re-read or verify contents later, but you want to continue with writing text at the moment. Here the common use of the marking "###" has proven to be practical. This marking can be found easily, using the function Edit—Find. Use this marking wherever you still need to look up or add details later. Besides, if you are not finally happy with your own formulations and want to rework the text later, it is useful to mark the text with the marking "###".

Coordination with the other group members
If you write your Technical Report in a team, the layout of the text, the structure and the terminology must be defined for all group members. If you do not follow this basic rule, inconsistencies may occur, which bother your readers or make it harder to read the report. Examples: different page numbering systems, different headers or footers, different methods to emphasize text elements, different part names in different chapters, different graphic design in drawings (usage of boxes, bars, lines, arrows, section lining, or area fill patterns).

To create a consistent report, the tools "style guide" (Sect. 2.6) and "report checklist" (Sect. 3.8.1) have proven their value. Especially during the initial stage of a project it is often hard, to use the established technical terms in a correct and consistent way. To assure a consistent use of terminology, include all technical terms which are important for your project (eventually with synonyms, which are also used in the literature with a short definition) into your style guide or into a terminology list. These lists or files are constantly updated with the progress of the project. If several group members are working on the report, the updates should be exchanged weekly or even more often.

Useful rules for writing the text

- If you take a larger break and switch off the computer, you should insert a marking before saving your currently edited file. This marking should occur only once at the position in your report where you have finished your work, e.g. "###break". If you open your file later, you can quickly find again the position with the function "Find".
- Always keep the latest version of the structure in sight (it is the "backbone" giving you guidance and support during formulating the text).
- When formulating and designing the Technical Report switch your mind to the position of the reader and look at all details from his point of view.
- Including figures and tables into your text usually improves the understandability, but it also takes a lot of time to create them. Therefore, create figures and tables parallel with writing your text. If you postpone this work too much, time problems are inevitable.
- If you want to bind a brochure together with your Technical Report, you should not write notes directly into the brochure. Better use note sheets which you lay into the brochure or Post-it sticky notes which you fix onto its pages.
- The pure typing of one page of the list of references takes two or three times as long as typing an ordinary text page. Therefore you should write the list of references parallel with the text into an own file.
- After a while, you will become routine-blinded against your own formulations. Therefore, give away your Technical Report for proofreading to a person who is technically educated, but does not know the details of your special project.
- If you are not very good in spelling and punctuation, you should also give away your Technical Report to a person who is good in spelling. Eventually buy yourself an Oxford manual of style (for British English) or a Chicago manual of style (for American English) or a similar book.
- Formulate the introduction and summary with special care. These two chapters are read thoroughly by nearly every reader!
- Plan enough time for the phases proofreading, master copy creation and end check. This makes your report more consistent and improves its quality.
- Always note things that still need to be done on a sheet of paper, in a to-do-list or a notebook! Otherwise, the risk to forget details is too high!

Structured working shortly before the dead-line

If you have found passages in your own texts, which need to be reworked, but you did not do this because of time-pressure, it is very probable, that your supervisor or customer complains about it. Therefore it is better, that you do not compromise with yourself and correct text passages you were not contempted with yourself immediately or insert the not-yet-ready marking "###" for still required corrections.

Sometime during the writing of your text, you should define a deadline, from when on no new literature sources shall be searched and read and no new information shall be inserted. This deadline should be at about 4/5 of the time for writing the Technical Report. If you find important sources after this deadline, you can still use them, but these should be only important articles or literature strongly recommended by the supervisor.

At the end check thoroughly, whether the report is "smooth" in terms of contents and layout. Check formal aspects with the report checklist and the style guide. Then your report will become consistent and reach a high level of quality.

After the written Technical Report is completed, you can approach the task to orally present your topic now.

4.7   References in This Chapter

http://dict.leo.org and
http://www.dict.cc dictionaries in various language directions.
www.systranbox.com/systran/box and.
http://www.google.de/language_tools?hl=en   translation   of   short   texts   in   various directions.

Presenting the Technical Report

5

Today the best Technical Report is only useful for someone, who can present it successfully. All what matters in the professional area—in doing business or politics—is strongly influenced by personal contact, by the spoken word, no matter how well it is prepared in written form. Therefore, if you want to have success, you cannot avoid presenting.

5.1 Introduction

The following pages introduce to you the characteristics, purpose, and background of presenting, taking a presentation of 20–60 min duration as an example. Then a systematic approach is shown that helps you to save time, money and nerves, when planning, creating and presenting your speech. At the end of this chapter, the characteristics of a short statement of 3.5–10 min duration are shown. 57 rhetoric tips from A-Z complete this chapter.

Without a systematic approach, just with a talent to talk it is impossible to present a good lecture or technical presentation. Please keep in mind, that during a presentation as well as when showing a piece of wizardry you can only come up with things "like that", which you have prepared before!

5.1.1 Target Areas University and Industrial Practice

You are a student just before presenting your master thesis or a postgraduate just before presenting your dissertation? You are an engineer and want to present a paper in your company or at a conference? If yes, the following chapter helps you to learn or improve

© Springer-Verlag GmbH Germany, part of Springer Nature 2019
H. Hering, *How to Write Technical Reports*,
https://doi.org/10.1007/978-3-662-58107-0_5

presenting your topic(s) to important people like your professor, your management, or a public audience.

The following rules, hints, and tips refer to examples from universities like student research project reports, bachelor or master theses or dissertations, but they are also applicable e.g. for an oral status report in a meeting in a company or a presentation of a new product at a fair.

Both roles demand similar work steps for a presentation
Student or engineer in practice—Do you want to slip into both roles?

Then you can probably agree with the following: What you get to know in this part of the book is valid for everyday life in university as well as for the professional practice. In all fields of life you are presenting your message to an audience of human beings, and they share the same expectations and behavior to a larger extent than usually estimated. You don't believe that? Try it out!

5.1.2 What Is It All About?

Initial situation
"My Technical Report is finally ready—presenting it won't be a problem!"
Or will it be one?
No, the lecture or presentation will not be problems, at least none that are impossible to solve!
Since you, dear readers, have created so many texts, tables, diagrams, and images throughout a long and busy period of time during the work on your Technical Report, that all this material should be sufficient for a lecture or presentation of 20, 30, 45, or 60 min.

However, is that the type of information you can quite easily create a good, successful lecture or technical presentation from? Why do we present lectures at all, when everybody who is interested can read everything in your Technical Report? These and similar questions shall be discussed and answered in the following sections:

– Which right to exist, sense and purpose does presenting a Technical Report have in modern business life?
– What needs to be defined in advance, e.g.

  – which target group(s),
  – which time frame,
  – which devices?

– How do you plan the creation and presentation of your lecture?

- How do you structure your lecture, which amount of information and level of knowledge do you select?
- What is important to get a good grade, praise or acceptance after your presentation?

This leads us again to the question: Why do we present lectures at all? What is all the effort good for—everybody can read everything!? What are the advantages of preparing and presenting a lecture? The answer can be found in Sect. 5.2. However, at first we want to point out the advantages for you personally, and which basic approach should be applied and which inner attitude to presenting is useful.

5.1.3 What Is My Benefit?

It is very easy—by presenting your report

- you as a person have much more influence,
- you can explain the contents of your report better and show the advantages,
- you will have more success,
- you will have more fun.

Let us begin with the fun: All presenting people will reveal (if they are honest), that it is an exaltation of your own abilities, zest for life, and also of having a little power, to successfully get your message across to people. Naturally speaking, this feeling does not come up when you give your first presentation. This is prevented by the stage fright, all presenters have at first, and which does never disappear completely.

However, with growing routine you will get a feeling of contentedness and satisfaction, to "communicate" with an audience ("to exchange something with the audience and then to have it in common"). You will have fun to get a welcome change from your at times boring work-aday routine, which helps you develop yourself in terms of communication and knowledge, i.e. which provides you with personal and professional success.

With your presentation the contents becomes more vivid, because due to your personal appearance and your freely spoken explanations it gets a much more intensive impact. To express it in easier words, your report mostly addresses "the head", the brain, the intellect of your readers, while the presentation reaches "the gut", the emotions, and the unconscious in your audience. Only if there is a complete and complementary impression of emotion (first) and intellect (second), you will provide a holistic and impressive experience of your work and your person standing behind all that work. There is an insight in psychology: When the feeling says No, there are no facts and arguments, which are good enough to convince your audience. You have already worked on your report for quite a while, all this effort shall result in a good grade, more money in your purse, and better career chances! Wouldn't that be a success?

5.1.4 How Do I Proceed?

The basic attitude you should take when starting to work on your presentation is:

Target group orientation
"What does my audience need and wish to get?"

The presentation of your report is something completely else than just summarizing or even reading parts of your written Technical Report! It is clear that all the effort and highly qualified work of many days, weeks, months, or in case of dissertations of years cannot be condensed to the ridiculously short time of 20, 30, or 60 min. Therefore, you can only select a small fraction of your Technical Report, sort of a peek through the keyhole. But which peek do you select? For this decision, you need three important phases:

- **Phase 1**: In your mind, try to get as much distance from your report as possible.
- **Phase 2**: Slip into the position of your audience including professor(s), your boss and other experts and try to anticipate their existing knowledge, attitude and expectations.
- **Phase 3**: You try to meet these expectations of your audience as well and complete as possible.

You will ask yourself: "Is all that really necessary?", "How can I reach that and how large is my effort to succeed?"

You will find the answers to these questions on the following pages, together with basic rules, hints, tips and tricks, which will help you to avoid mistakes, save time and be successful. Become surprised, step in!

5.2 Why Presentations?

In this section the properties of a lecture or a presentation shall be pointed out and the differences between lectures or presentations and Technical Reports. In addition, presentation targets and presentation types are described. However, at first we want to introduce a few definitions of terms, which are often mismatched.

5.2.1 Definitions

- **to present**: introduce, show, make clear and present
  (present as a gift, present as present tense)
- **lecture**: longer oral speech (20–60 min) about a topic or on a certain occasion with transferring knowledge and/or influencing opinions
- **presentation**: lecture of medium length (10–15 min) about a project or a product with transferring knowledge and influencing opinions.

- **statement**: short lecture (3, 5–10 min) on a topic, for example expertise, point of view, own opinion.
- **information**:

  1. shortest lecture just for informing other persons
  2. other word for transferring knowledge (should always be objective and neutral!)

Naturally speaking these terms cannot be divided from each other very sharply, but they have common properties. The lecture is the master type of all speech forms and it is most elaborate to prepare and to present. All other speech forms can be derived from the lecture and have more or less elements in common. Therefore, we should get to know the properties of a lecture especially well, even if it is abbreviated in a presentation or a statement. Therefore, if we speak of a lecture, we also want to address the derived speech forms as a whole or in parts.

5.2.2   Presentation Types and Presentation Targets

Before we speak about presentation types and presentation targets we want to explain the term "to present" in three aspects: First, to get across the contents of a lecture, i.e. to make it present for the audience, a good speaker should second be present as a person, e.g. via charisma and engagement, and third he should give his lecture to his audience as a present. Presents are exchanged between friends. Therefore, at least during the lecture and the following discussion, there should be a basic attitude of friendship between speaker and audience (a friendly relationship aspect). We want to define four important presentation types, Table 5.1.

The transitions between these presentation types are fluent!

Presentation targets A lecture, presentation or speech always has three targets, each with a different weight:

1. **Informing**: Technical knowledge

   - shall be documented (saved in texts, figures and images) and
   - transferred (communicated to other people)

2. **Persuading**: Persuading the audience

   - of the quality of the provided knowledge
   - the efficiency of the delivered work
   - the competence of the candidates and/or co-workers involved

Table 5.1 Presentation types

Presentation type	Properties
1. Lecture	– The preference is pure information! – Objectivity is a must – Main contents: technology, presented appealingly – The audience is addressed so that they understand the contents easily, i.e. indirectly
2. Persuasive presentation	– The preference is persuasion with technical and non-technical arguments – Technical objectivity only as far as it is needed – The presentation must be especially appealing, persuasive, and eventually amusing – The audience is addressed very specifically and directly
3. Technical presentation	– The preference in most cases is influencing the audience – Still a lot of technical information is provided
4. Occasional speech	– The preference is influencing and amusing the audience and appealing to their emotions – Only a little technical information is provided (minor point)

3. **Influencing**: Influencing the audience to act:

   – granting money
   – buying
   – continuing the project
   – positive decision
   – good grade

5.2.3 "Risks and Side Effects" of Presentations and Lectures

In all technical professions there are still written and oral forms of instruction and communication, where "written" includes the text-based communication via data networks. Why is none of these communication forms sufficient? The answer is: Both communication forms have advantages and disadvantages, comparable with the "risks and side effects" which you have often heard about. Let us look at both communication pathways with all their strengths and weaknesses, Table 5.2.

The written communication (documentation) comes up as notes, reports, manuals, electronic messages or written offers.

The oral communication contains arguing and presenting in a talk, a meeting, or in a presentation or lecture.

Both forms have about the same number of strengths and weaknesses. Therefore you need both forms in your professional life, and therefore this book does not only show you how to build up a Technical Report but also how to present it.

Table 5.2 Advantages and disadvantages of written and oral communication

	Advantages	Disadvantages
Written communication: documenting, i.e. from "note" to "offer", incl. e-mail	Independence of time, editing is possible at day and night: all contents is visible; "reader" can control his/her reading and read again: contents is clearly set; "writer" can work with concentration, paragraph-by-paragraph, take breaks; quality can be controlled well. Verification is always possible	Reader cannot ask back, impersonal, inflexible; no control of the effects and which reader reads under which circumstances, no feedback; What is written, can hardly be taken back/smoothed/emphasized
Oral communication: arguing, presenting, i.e. from "talk" to "lecture"	A good speaker is enjoyable for the audience; speaker has contact, gets feedback, can make convincing use of his personality; can correct himself, can react flexible, can make use of/control moods and emotions (up to demagogy and manipulation); spoken word, visible images and physical presence of the speaker altogether result in a strong impression	Missing reproducibility, need for physical/psychological strength, strong nerves, active knowledge, discipline, intellect, quick-wittedness, knowledge of people, sensibility, eloquence, good manners, positive impression, charisma, good form on that day! Disturbances because of comments or questions or intentional interferences; possibility of misunderstandings, risk of errors in reasoning is always there, constant time-pressure, stage fright, self-doubt

The question asked at the beginning "Why presentations?" can be answered now:

▶ To convince people of technical contents and to influence them in the desired way you need more than written communication. Human beings are no "scanners" (which would be sufficient for the pure perception of the information contained in your report), but as decision-makers they are also characters with head and gut. In addition to reading your report with the "scanner function" the senses "Hearing" and "Seeing" want to be addressed. Last but not least, the intuitive feeling of your personality via voice and visual impression is also essential to convince and influence your audience as far as possible.

Voice and images are the basic elements and strengths of a lecture; they make the oral presentation inevitable, if there is something important to decide. That is why speeches in parliament, oral lessons at school and in vocational education as well as lectures in study courses play a dominant role. Finally another, important aspect: The personality of the speaker is not only transported via the voice, but also via nonverbal signals like erectness, facial expression, gestures and charisma. They have a strong influence on the audience, which is missing to any report or book. Therefore, when standing behind the speaker's desk we should know these relations, processes and results and build up our lecture completely different from our Technical Report.

▶   The lecture or presentation is a new creation—based on the report, but with completely different focal points, images and elements of style—if it shall be successful.

Now you have enough background knowledge and motivation to plan your presentation.

5.3 Planning the Presentation

Yes, a creative chaos has its sympathetic sides, but without planning you will soon run into time-pressure with respect to the date of your presentation. Therefore, the following pages describe the work steps to prepare and present a lecture and their time consumption.

5.3.1 Required Work Steps and Their Time Consumption

Figure 5.1 shows a network plan with the required work steps ordered by their time-sequence. This plan recommends seven preparative steps, which need quite a different amount of time before in the eighth step the presentation is performed. These tasks do not run strictly one after the other. Please refer to the Gantt diagram, Fig. 5.2. All tasks and recommendations for your time planning refer to the essential presentation, which needs to be as perfect and successful as possible, like presenting your master thesis or applying for a job or project continuation.

Naturally speaking, "normal" lectures created in day-to-day business can be created with more freedom and less time consumption as your routine grows. The proposed time frame assumes that you have available two workweeks with two weekends. That means the preparation of your lecture takes place partly in the spare time during the week and partly during the weekends.

The weekends are mainly needed for step 3 "the creative phase: gain distance! develop a structure" and for step 6 "trial presentation, changes". Especially for step 3 you will need some calm and lonesomeness to unfold your creativity, which is not provided at most workplaces. The Gantt diagram (Fig. 5.2) and Table 5.3 show recommendations for the required time you should take into consideration.

The tasks described above (Steps 1–8) are explained in detail in the next sections.

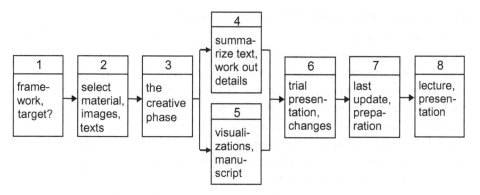

Fig. 5.1 Network plan to create and present a lecture

Worksteps	Time consumption	1st week	2nd week	1 – 2 days
1	define presentation framework and target	▪		
2	collect/select or create material, images and texts	▪▪		
3	The creative phase: gain distance, develop a structure		▪▪	
4	summarize the text, work out the details		▪▪	
5	create (more) visualizations, write the manuscript		▪▪▪	
6	trial presentation, include changes			▪▪
7	last update of the lecture, preparations in the room			▪
8	hold lecture or presentation			▪

Fig. 5.2 Gantt diagram to create and present a lecture

5.3.2 Step 1: Defining the Presentation Framework and Target

Planning your presentation consists of step 1 "define presentation framework and target" and step 2 "material collection". With step 1 you build a solid foundation for your presentation by defining all important presentation conditions, the target group and the information targets of your lecture before executing the next work steps. This helps to prevent a failure and to save time. You should ask for the following presentation conditions and targets, discuss them in your team or define them for yourself:

Table 5.3 Time consumption to create and present a lecture

Steps in the network plan		Time consumption
Step 1	Define presentation framework and target	1 day
Step 2	Collect/select or create material, images and texts	3–5 days
Step 3	The creative phase: gain distance! Develop a structure	2–3 days
Step 4	Summarize the text, work out the details	2–3 days
Step 5	Create (more) visualizations, write the manuscript	3–4 days
Step 6	Trial presentation, include changes	1–2 days
Step 7	Last update of the lecture, preparations in the room	1–2 h
Step 8	Hold lecture or presentation	20–60 min

> **Presentation framework (presentation conditions, target group, information target)**
>
> – What do I talk about? (topic)
> – What shall be the title of my lecture?
> – What kind of lecture will I present? (presentation type)
> – Who will listen to me? (audience, target group)
> – What is the occasion to present my lecture?
> – What do I want to achieve with my lecture? (presentation targets)
> – Why do I want to achieve that? (purpose)
> – Where do I present my lecture? (surroundings, room)

Clarify these questions in detail (if possible in written form). Then you will feel a little safer, and the lecture cannot be a complete failure any more. In the following, we want to look at these questions a little closer.

What do I talk about? (topic)
Very easy—about your Technical Report!

However, the title of your lecture may differ from the title of your Technical Report, e.g. if the latter is too long (occurs often) or if it sounds too complicated for the title of a lecture or if a better title came to your mind.

In professional life you will hold lectures about various, important topics. Then the topic and contents should be agreed upon as exactly as possible with your customer, or boss or chairperson of a conference or organizing staff of a fair. This includes a (written) definition of

– title, subtitle
– audience (knowledge level, interests, expectations)
– exact timeframe without/with discussion

– topic and presenter of the previous and next lecture
– presentation type, room size, equipment (desks/chairs/boards), devices (projectors/flipchart).

For your career, it is very important to arrange things with

– your boss
– your supervisor, professor or customer or the organizing staff of a congress, conference or fair
– your own sales and marketing team leader!

Why should you speak with the sales team leader and the marketing team in any case?

Presentation framework in the automotive industry
Your company, a well-known car manufacturer, has developed a super-fast sport coupé and wants to introduce this marvel with an expensive PR show at the next fair. 14 days before the fair you (being a young, enthusiastic development engineer) present the most important motor and chassis details on a conference. Among the experts in the audience is a representative of the press who launches your insider knowledge big on the next day. Which consequences would that have for your company, for you and your job?

Now let us go back to the presentation framework, that needs to be clarified.

What kind of lecture will I present? (presentation type)
Is a pure technical presentation or a special persuasive presentation more appropriate? Does the audience expect pure technical facts of your report or do they want to get insights they did not have before? Do they want an overview or details? How do you mix facts, influencing and emotion best, maybe even with a pinch of humor? This depends on your own direction and information target, but also on the audience and the occasion to present your lecture. Therefore, you should also clarify the following four questions.

Who will listen to me? (audience, target group)
Are they your professors or supervisors? Are they engineers, bankers or journalists, are they experts or unknown visitors, are they your boss and other executives or are they colleagues from sister departments? You should best prepare your lecture depending on these people or—if you do not have better prior information—for a colorful mixture of people. It will be explained how to do that with respect to the aspects "amount of information" and "knowledge level" (see Tables 5.4 and 5.5 as well as Fig. 5.5).

What is the occasion to present my lecture?
Will you present your master thesis or your dissertation? Is it a pure report for colleagues or a presentation for customers? Do you want to persuade investors (banks, scientific funding boards, federal ministries) or a working committee for standardization or judges in a trial? You will have to adopt your lecture to these occasions more or less, if it shall be successful—and that is why you present it!

What do I want to achieve with my lecture? (presentation targets)
This question deals with your presentation target(s). This means, whether and in which mixture you want to achieve the following effects:

- Transfer of technical and scientific knowledge (to make your audience cleverer), or
- creating a positive impression (of your technological or scientific approach, of you or your company), or
- creating emotions in your audience (in a desired direction).

Naturally speaking, you do not follow these information targets too obviously, but smart and carefully—Your audience should not realize your tactic, at least not consciously. As long as your approach does not exceed to crude manipulation, no one will complain about a decent rhetorical tactic.

Why do I want to achieve that? (purpose)
Here you have to define the purpose of your presentation.

Purpose of the presentation
Is everything allowed that advances the purpose???
Let us be careful and say: Sometimes yes!
An example:
The future funding of your whole project group depends on your lecture. It is a "To be or not to be"-affair for the jobs of your colleagues and maybe for your job as well—isn't that worth to invest a lot of effort?
Such purposes in mind, you will probably be willing to pull out all legal stops to be successful, won't you?

This includes deciding how much you want to convince, influence or amuse. These elements of a good lecture must be mixed well and used with care. That requires a good preparation, because mixing these targets ad hoc does only work for a few routiniers. Preparation means to think about, define and test things.

Where do I present my lecture? (surroundings, room)
Print out a description how to reach the presentation room and clarify room capacity, desk layout, needed media (board, flipchart, overhead projector, computer and beamer, moderation kit, felt pens) as well as eventually catering for the audience and book the room, if necessary.

To explain the step 1 "define presentation framework and target" and the consecutive steps 2–8 more concrete, an example shall accompany us from now on, as far as it is possible to describe it within the framework of this book.

Note: The student's research project, the following example presentation is derived from, is listed in the list of references. The conference and the lecture did never take part. The "Institute for Welding Technology" does not exist at Hannover University. The workflow is just invented to describe the basic approach to you. The example presentation will be in italic letters from now on.

The title of the Technical Report in our example is:

Title of the Technical Report

*Improving health and safety at work when welding
by using burner-integrated suction nozzles –
Effectivity and quality assurance*
Student research project at Hannover University

The Technical Report that describes the student research project consists of 135 pages and shall be presented in 20 min. How can this be done?

The author courageously starts and clarifies in step 1:

- **Topic**: The student research project named above shall be presented as a paper on a conference about welding technology.
- **Title**: "Burner-integrated suction when welding"
- This title is clearer, shorter and better memorable than the title of the report, thus more attractive and still correct.
- **Audience**: Head of the institute (professor), supervisor, fellow-students, a person working in industry, a journalist, other participants with unknown technical background.
  This audience has various knowledge levels, interests and expectations, which must be taken into consideration skillfully. Nobody should be overstrained or bored!
- **Timeframe**: 20 min lecture with additional discussion.
- **Previous lecture**: "Health and safety in crafts" (Person from the employers' liability insurance association)
- **Next lecture**: "Automated flame cutting" (Person from the company Messer-Griesheim)
  To avoid interferences and contradictions, it is necessary to contact these presenters and to harmonize the contents of the three lectures!
- **Presentation type**: Technical presentation
- **Room**: Lecture room 32, 30 seats, desks
- **Media**: Overhead projector and beamer available
- **Agreement with professor and supervisor**: Planning discussion: Rough selection of the contents, strategy to meet the persons from industry and the journalists

Building up on this presentation framework the speaker designs her plan of action. She especially thinks about the mixture of technical information, convincing and influencing taking into consideration the following criteria:

- **Target group**: "The audience described above is very heterogeneous. This requires creating a balance between experts (professor, supervisor) and the other listeners."
  This balance can be created with a smart selection of the contents (see Sect. 5.3.4).

- **Occasion**: Seminar or conference "Current state of welding technology", Obligatory for students, public event. Backgrounds are also to gain funding from industry for the institute and to improve the image of the institute.
- **Targets**: The lecture shall mainly serve to instruct the audience about the findings of the project (70%), but it shall also show the competence of the institute (20%) and introduce ease/humor into the uninspiring contents (10%).
- **Purpose**: Main purpose: Good grade!
  Desired side effects:

  - Awake interest in industry and crafts.
  - Get a good echo in scientific journals.
  - Bring knowledge and motivation to the students.

Step 1 is now completed. Let us continue with step 2 "Material collection".

5.3.3 Step 2: Material Collection

To work on step 2 two situations must be distinguished:

- **Case 1**: The material (the contents) of the planned presentation already exists, e.g. in the Technical Report.
- **Case 2**: The material must be collected, found, created.

In Case 1 we have the complete report available. Then it is the task to select the information suited for the presentation from the rest of the information contained in the report. Suited means the information is essential, meaningful and representative. Thus, in Case 1 the material collection mainly consists of the selection of contents. Please refer to Sect. 5.3.4 for more details.

In Case 2 we do not have a ready report. This is more difficult. The material must be collected or found, e.g. in textbooks and binders in your office, in archives, libraries, or the internet. This may take about 3–5 days. If this time is not sufficient, you are not enough an expert and cannot present this topic convincingly. Someone else should hold the lecture.

If you have found and printed or copied enough material, you should sort out the less important material (texts and images) and put it aside as a reserve.

The other material should be sorted by:

- literature sources (consecutively numbered),
- topics
- and/or author(s).

To do this systematically you should create a file/an index.
This completes the routine work for your presentation.

5.3.4 Step 3: The Creative Phase

In step 3 the most important, most difficult, and most exciting phase on your way to a successful presentation takes place:

- Gaining distance to the contents, gaining an overview from the perspective of the target group (audience) and
- finding (creating) a structure, which represents the relevant contents and meets the expectations of the audience.

That is no easy trick. In spite of the fact, that you are an expert in your topic and that you have worked out and documented it, in your mind you step next to yourself and imagine the situation of your listeners. Meeting their expectations you develop a concept for your presentation, a basic outline of what you want to describe, maybe with a special gag and with a dramaturgy—as much science as needed, as much clearness as possible ...

How far you (may) deviate from your Technical Report and its structure only depends on the targets and purpose of your presentation or lecture. This may result in a completely new structure of the information, which strongly influences the lecture and makes it successful.

This creative phase may last 2–3 days—it makes your lecture special. Let us look again at the speaker in the example "Burner-integrated suction when welding". She takes her readily written project report and analyses its structure.

Structure of the Technical Report	
1. Introduction	(2 pages)
2. Fundamentals (Welding methods, pollutants and their removal)	(29 pages)
3. Optimization of the burner-integrated suction	(24 pages)
4. Welding tests with the optimized suction	(68 pages)
5. Summary and outlook	(4 pages)

This structure of a Technical Report is sound and informative, but not well suited as a structure of a lecture or presentation. Why?

A structure of a presentation needs a few, short, inspiring items, which can be memorized well and have a dramaturgy. Our speaker creates the following structure of the presentation.

Structure of the presentation	
Introduction	Health versus seam quality? ("wow effect")
Main part	– Suction techniques (current state)
	– Burner-integrated suction
	– Design optimization
	– Tests to measure the weld smoke (measuring the isotachs)
	– Tests of the seam quality (destructive and non-destructive)
Conclusion	Summary, answer to the "wow effect"

A Technical Report or another technical topic can be transported like this or in a similar way in 20 min. Naturally speaking, such a concept should not result in a sales presentation. It depends on your instinctive feeling here not only to convince the professor who will give you a grade by an obviously serious scientific approach, but also to raise the knowledge and create a positive attitude towards your topic in all the other people in the audience by an understandable and clear approach and your personal charisma.

Realizing the structure
Now it is necessary to realize the structure, i.e. to make it vivid with statements and information, texts and images. Everybody in your audience from the professor or experts to the accidental guest shall "take something home" from your presentation—all of them shall think that your presentation was a present, at least partly.

Can this task be fulfilled? Yes, it can—partly! Partly means: Everybody in your audience will get his part of the contents of your presentation, but not all the time. The secret is to smartly distribute the contents of your lecture depending on the individual expectations of the different persons or groups in your audience:

– Widely known facts for all listeners ("Well-known")
– Complex scientific information for the experts ("Mazy")
– Understandable new information for the majority ("New")

The third-rule
As a presenter, you will seldom have an ideally homogenous audience, because even among experts there are big differences in their knowledge the more special our modern knowledge becomes. In addition, you want to prepare your presentation only once and re-use it for more than one occasion, if possible. Therefore, you should apply the third-rule, Table 5.4.

The third-rule may look a little tricky, but it is the best way to find the clue a very inhomogeneous audience needs. It has often proven to be practical—find your own judgment, try it out!

To realize the third-rule you cannot just present the parts one after the other according to the fractions, it is a little more differentiated. A presentation should not be split into three exact thirds, but it has three parts.

Table 5.4 Third-rule (trisection) to design a presentation

Part	Explanation	Purpose
1/3 "Well-known"	Directly affecting, daily things, everybody knows	Common base, self-affirmation to the audience
1/3 "Mazy"	Special facts and details, which are only understan-dable for the experts	Raise of knowledge for the experts (insiders), image improvement for the speaker
1/3 "New"	Scientific and/or technical information on an appropriate knowledge level, understandable for the majority	Raise of knowledge for the majority, eventually even for people who are not acquainted with the subject area

Realizing and using the trisection

This trisection can be found in the biology of daily life and in technology, as the following examples show, Figs. 5.3 and 5.4.

This trisection is realized in a technical presentation by smartly varying the knowledge level, to meet the individual expectations of as many people in the audience as possible, Table 5.5.

Admitted—the exact distribution of the main part seems to be a little artificial and farfetched, but it makes sense as an orientation and recommendation, especially due to the section into thirds or sixths resp. according to the third-rule. What is the secret?

After contact preparations and contacting the audience, which are described in Sect. 5.5.1, the presentation should start with the phases "presentation target and course of action" and "introductory examples" to give an overview and build-up an information base for everybody in the audience. This shall activate ("warm-up") the audience and raise their interest for the topic of the presentation.

Then scientific and/or technological information on an appropriate knowledge level should be delivered, which is understandable for the majority of the audience—the first sixth of the "New". Before the experts in the audience become bored, you refine the

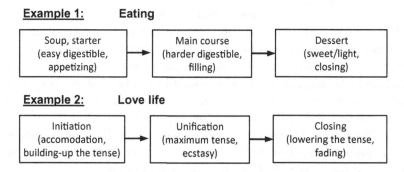

Fig. 5.3 Processes compared with a presentation (biology)

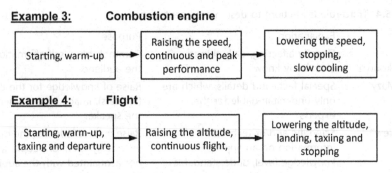

Example 3: **Combustion engine**

Example 4: **Flight**

Fig. 5.4 Processes compared with a presentation (technology)

Table 5.5 Design of a technical presentation

Phase	Time	Contents	Information target and fraction according to third-rule
Introduction	About 20%	Contact preparations and contacting the audience, presentation target and course of action	Warm-up
Main part	About 70%	Introductory examples, overview, information base	1/3 Well-known
		Raise of knowledge	1/6 New
		Refinement, details, specialties	1/3 Mazy
		Aggregation, conclusions and evaluation	1/6 New
Conclusion	About 10%	Summary (outlook)	Closing

information and present the "Mazy", i.e. more complex connections, details and specialties, which can be understood only by one or a few experts e.g. your professor and his colleagues. However, in this part of the presentation even these experts should be able to still learn something.

During the phase "Mazy" the non-experts, i.e. the majority of the audience have growing difficulties to follow the presentation. They stay in respectful amazement of the knowledge and brilliance of the speaker or partly turn off their conscience. Before the high knowledge level starts to completely frustrate your audience, you go back to an appropriate knowledge level and present the last sixth "aggregation, conclusions and evaluation". Now the majority can follow your presentation again. Here the non-experts also raise their knowledge and insight, so that the presentation was worth listening for them, too.

A short, but meaningful summary of the most important findings and key aspects of the contents of the presentation, eventually a short outlook, but no new, essential information and friendly sentences form the Conclusion of the presentation.

This juggling with three balls with respect to the knowledge level is the secret of a good speaker, if he/she wants to satisfy all members of his audience. Everybody in the audience shall have learned something, what he/she can write into his/her diary. No one should think or write: "I knew all that before" or "All that stuff was too complicated"—This would result in: "Much ado about nothing!".

In a graphic model, this just described design of a technical presentation could look like it is shown in Fig. 5.5.

In this dramaturgy, everybody in the audience has a benefit from listening to your presentation. After a quick, steady increase of the knowledge level the majority learns something new, and then the experts are informed on a high or very high knowledge level. In this phase, the knowledge level peaks return frequently to the level of the majority, to keep up the conscience of the non-experts and guarantee transparency. Shortly before the end of the main part the knowledge level is quite high.

Everybody shall take something valuable home. That is also important for the usually following discussion, which is useless for the non-experts without a certain level of understanding.

The speaker in our example "Burner-integrated suction when welding" performs the trisection for her presentation, Table 5.6.

In Sect. 5.4 you will find more details how this theoretical model can be realized in practice. Especially the inevitable trial presentation will relentlessly show what can be done and what cannot be done in 20 min.

5.4 Creating the Presentation

The concept is ready—now it must be elaborated! The structure with its skeleton for the contents and the time consumption must be transformed to a clear, successful presentation.

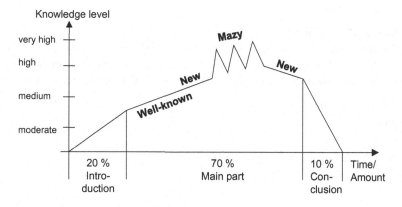

Fig. 5.5 Dramaturgy model of the knowledge level

Table 5.6 Trisection and assignment of contents in the example presentation

Main part (with time planning)	Phases (with time planning)	Contents
4 min Introduction	Contacting the audience, pre-sentation target, overview	– Is the seam quality more important than health and safety at work? (provocative question, raises tense) – Examples of the problem (clearly described)

14 min Main part	4 min "Well-known" (state-of-the-art)	– Removal of pollutants during inert gas welding
		  – Example suction in a working hall
		  – Example suction at a workplace
		  – Example suction with a hood near the burner

	2 min "New" (improvements)	– Burner-integrated suction (process, parts, problems) – Optimization of the VACUMIG welding pistol (suction geometry, seam quality)
	5 min "Mazy" (special know-ledge)	– Suction tests (test details, combination of suction rings, test results) – Welding tests (test details, program plan, example of measuring protocol, test results)
	3 min "New" (consequences)	– Evaluation of suction tests – Evaluation of welding tests – Error calculation – Technical consequences
2 min Conclusion	Summary (outlook)	– Summary of results and findings – Future work

5.4.1 General Recommendations for Designing Presentation Slides

In this section you will find general hints for designing presentation slides. This book can only show the general direction. You can find more details in the online help and in the appropriate literature.

Basic layout of the slides
At first you should select the basic settings for the slide and title master. Here you can influence the layout of the list or text object resp., which is contained in the preset slide layouts List, Two-column text, Text, Text and ClipArt etc. On the slide master you can select

– font type, color and size,
– language,
– tabs,
– form and color of the bullets in lists, and
– line spacing.

In any case, you should move the left margin of the text area and the title area, to gain space for your structure. Now you can create structuring lines or other graphic objects. To modify the background, you can select the background color, images, area fill and the logo of your company, institute or department.

ou recognize later, when you create your presentation slides, that something basic is (still) wrong with the text fields or other systematically repeated elements, you should preferably change that on the master slide.

Font type and size
It has proven to be practical to use fonts without serifs like Arial. The font size should not be smaller than 18 points, the larger the better.

Use of color
Colors can occur as text color, background color and object color in images. If used with care, colors help to specify, emphasize, structure, order and differentiate. Colors control reactions, stimulate imagination and initiate emotions, memories and associations. Use the same colors for the same things and use them consistently! If colors are used sparsely, they serve as "attraction" and help to emphasize important details.

However, color reduces the contrast between background and foreground and the recognizability. You can recognize best black on white. Text written in dark blue, dark green and dark red on white must already have a 1.5 times larger font size and lines must be 1.5 times thicker to achieve the same recognizability. Too many colors without explanation foster the stimulus satiation and burden the overall impression.

Readability check
To check the readability you can show a critical slide in screen presentation mode. Take a folding rule and measure the width of your screen, multiply this value by 6 or according to DIN 19045 even by 8, and then stand in front of your screen at that distance. Alternatively print out this critical slide on DIN A4 paper, lay the printout onto the floor and look at it while standing upright. If you can read everything, it is OK.

Presentation framework slides
It has proven to be practical, that every presentation has three framework slides to assure transparency for your listeners:

- The title slide (presentation title, name of the speaker, name of the institute, company or department),
- a structure slide, showing the structure with normally formulated, long items (can be replaced by a structure written onto a flipchart sheet or a poster at the wall) and
- the termination slide (contains a graphic or a photo to encourage the discussion and shows the contact data of the speaker).

Transparency assurance
Transparency assurance is one way to consecutively display the current items in the structure like introduction, main part with subordinated items and summary/conclusion on the slides, Fig. 5.6, and thus assuring the transparency of the presentation.

A modern, self-confident audience always wants to know, where they are in the flow of the presentation. Then they can better prepare for the current aspect of the topic and do not start floundering through your presentation.

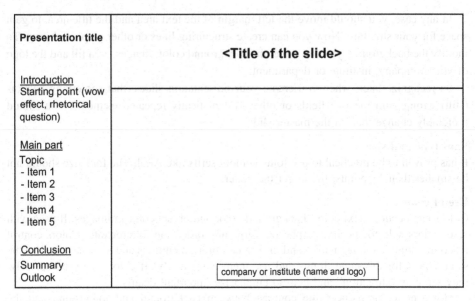

Fig. 5.6 Transparency assurance by a structure in a sidebar (schematic)

The structure of the presentation should be always visible to your audience to provide orientation. The speaker should clearly point out, where he/she is within this structure from time to time. This can be achieved by

- repeated reference to the structure and depicting the current position by ticking off the finished items at the flipchart, on a poster or a (white-/black-)board or
- by a structure in a sidebar (= area on the left or at the top, max. ¼ of the total slide area, separated from the visualization, showing the preferably very compact structure, having selected the current item). This is the only option, if in the lecture room there are no boards, no flipchart and no possibility to fix a poster to a wall.
- The structure can also be handed out to the audience prior to the presentation.

The prerequisite for this method of transparency assurance by a structure in a sidebar is a very compact structure with short keywords!

Slide design, entering text, including images
For slide design keep in mind the general principle "The fewer, the better!". Avoid everything on your slides and in your beamer presentations, which is dispensable and can lower the level of conscience of your audience. For example, avoid the:

- constant repetition of the presentation title,
- constant repetition of your name and the name of the institution,
- constant repetition of the date and occasion,
- constant repetition of copyright notes,
- constant repetition of your personal slide management notes (e.g. "Slide 8 of 31") etc.

Does the "constant repetition" bother you as well?

Admitted, this can easily be created with the computer, but the audience—as curious as it is—will (have to) read all these things again and again, just to check whether or not something has changed …

To enter text you should use the predefined text object area with the bullets. To change the indentation of a list you can use the buttons $\Rightarrow$ and $\Leftarrow$ or mark a point in the list and press Tab or Shift + Tab to reverse-indent.

Another useful option, which only works, if you use the predefined object areas, is the fast entry of text slides in the structure view. By typing Enter, you create a new slide. If you enter text now, this will create the slide title. Press Enter again. A second new slide appears. Select this slide and indent it to the right with the right arrow key. Now you can enter text for your first slide. Subordinated items are indented with the right arrow key again.

There are many possible methods to insert images, which can only be described rudimentally here. For example, right click a slide and in the context menu select Slide layouts. Select the layout Text and ClipArt, enter your text on the left and insert a ClipArt graphic on the right.

If you want to cut an image, you can use the cutting tool. This makes parts of the image invisible, but they still exist.

Entering the structure at the left margin of the slides

The structure at the left margin of the slide can best be created as a text field. Insert the field, enter your structure, format the text as desired and move the object to the desired position. If you click into the text field, it is selected. If you click on the border now, you can position your text field with the arrow keys. An even more exact positioning is possible, if you press the Ctrl key, keep it pressed and then use the arrow keys. If you clicked the edge of the text field, you can copy copy it on one slide with Ctrl-C and insert it into the next slide at the same position with Ctrl-V. In addition, you can delete it with the Del key.

Footers

You can often see, that presentations have footers containing the presentation title, the speaker's name and a page numbering, like "8" or in extreme cases "Slide 8 of 23". Your audience has to re-read that on every slide. This reading of irrelevant information uses up a part of the brain's capacity, which is not available any more for understanding the important contents of the presentation and it needs precious time. In addition, this just bothers your audience!

That means, that due to the reasons mentioned above the footers of slides should contain as little information as possible. If you show your presentation with a beamer, the page numbers are superfluous. They are only helpful, if you show your slides with the overhead projector. But there is no rule without exception: For advertising and prestige purposes and due to juristical reasons many institutes and companies have corporate identity (CI) rules, which define a certain framework for all their presentation slides. This includes showing the name and logo of the company. We did that in our presentation example as well. Copyright notes may be necessary to preserve the company's copyright. However, this occurs seldom. Sometimes after filling up the slide with formalities like this, there are only 50% or less of the total slide area available for the actually interesting information. Try to minimize or eliminate this kind of side information as far as possible. A more attentive audience will thank you.

However, for notes and handouts you should use footers (menu View—Headers and footers, Notes and handouts tab), because e.g. with page numbers you will facilitate the paperwork for your audience!

But now let us go back to elaborating your presentation.

5.4.2 Step 4: Summarizing the Text and Working Out the Details

Step 4 is the practical realization of the details planned so far. What belongs into the 4 min Introduction? In general the presentation target and the course of action in the presentation or lecture. In addition, a demarcation is made, which aspects of the topic the speaker cannot or does not want to deal with. The course of action can best be explained by means of the projected structure slide, which may be handed-out to the audience (the first present!).

In Sect. 5.3.4 we have already discussed how the structure of a presentation or lecture is build up (in contrast with the structure of the report described in Sect. 2.4.4). Table 5.7 shows a structure design plot, which will later be used for the example presentation "Burner-integrated suction when welding".

It is important that you leave out everything which is unnecessary on the slides: No "Speaker", no "Presentation of …", no "Contents", no "Structure", no "Welcome", no "Introduction", no "Closing words", no "Thank you"! This structure shall not represent all the contents of the presentation—it shall just give a rough orientation for the audience.

The audience will quickly get used to this short and dense presentation structure, the speaker can have this structure completely present in his/her mind, and always in sight, what makes your presentation appear more complete. The graphic design of this structure can vary as much as you like—as long as it is not overburdened. The few, a little vague items of the structure also have the advantage that they do not determine the speaker too

Table 5.7 Structure design plot for 20–60 min presentation time

Structure design plot		20 min presentation	60 min lecture
Title	Clear, short, simplifying, but still accurate, memorizable!		
Speaker	(Title) first, (middle) and last name, place (city or address of the company)		
Introduction	Informative, inspiring, demarcating!	4 min	8 min
Main part	Clearly structured, not too many sub items, easy to follow, logical, consequent, fascinating, dramaturgy-oriented!	14 min	48 min
Conclusion	summarizing, harmonizing, nothing new!	2 min	4 min

much—the audience will check whether all items have been covered in the presentation without mercy.

Now this structure design plot will be used for the example "Burner-integrated suction when welding".

Titel:	Burner-integrated suction when welding
Vortragende(r):	Franziska Benz, Hannover
Introduction:	Health versus seam quality?
Main part:	Suction of welding pollutants
	– burner-integrated
	– optimized design
	– verified quality
Conclusion:	Summary, outlook

If the author and the speaker are different persons, this must be specified!

The further summarizing of the text shall cover exactly this framework and all changes to the next item in the structure shall be noticeable for the audience. Items, which have been completely covered, can be ticked off at the flipchart, the current item in the structure can be emphasized in a structure at the left margin. A presentation made transparent like this gives more security to the speaker and to the audience, the speaker is always the "master of the situation".

Working out the details contains the selection of the exactly suited facts, figures and statements for each item in the structure. Restrict yourself to the essential information and leave out all excurses, as interesting as they are for you! Parallel with this selection, you create (more) visualizations, the text slides and the manuscript (Step 5).

If you have turned back the pages, you have probably noticed the discrepancy between the exact, detailed backbone of the example student research project thesis and the short, nearly superficial structure of the presentation. That is desired—the speaker has planned to show a lot, but she does not show all her cards and thus remains more independent and flexible.

The preparation of the presentation contains the important Step 5: Visualization and manuscript, whose special properties will be described in the next section.

5.4.3 Step 5: Visualization and Manuscript

Technical presentations and lectures cannot be presented without images (visualizations). This statement is not only true for presentations, but also for the Technical Report, whose visualization is described in detail in Sect. 3.4. However, for presentations there are a few more rules and recommendations, which will now be introduced.

In contrast with readers of a Technical Report, who can look at each visualization under individually selected conditions, look at them as long, intensive, and detailed as they like and study the explaining texts, an audience can only consume what is shown by the speaker during the presentation in a passive mode. And the audience shall follow the speech and compute the provided information. The following overview shows the influence factors on the visualization of the presentation and their main properties. The visualization shall be "striking". What effect does a striking design have? Which properties does a good poster have?

- It is appealing.
- It acts fast.
- It transfers simple messages.

Visualization of the presentation
The four main differences between visualizing a report and a presentation are:

- **Time factor**

  The time provided for looking at the individual visualizations is only determined by the speaker, and in most cases it is quite short.

- **Reproduction quality**

  It is logically lower because of the necessary projector or beamer and projection wall.

- **Larger distance**

  Compared with the normal reading distance of 30–40 cm the distance in a presentation is 10 m and more.

- **Disturbing influences**

  They derive from other people in the audience and room factors like too bright light, glare or obstacles to look at the projection wall, e.g. the speaker or an arm of the projector can trouble your audience.
  These conditions require a presentation-suited visualization with the following properties:

- **Minimum contents**

  Restriction of the message of the image to the absolute minimum, no sentences or even paragraphs to read, no complex illustrations.

> **– Striking design**
>
> Emphasized, eventually exaggerated layout with thick lines, simplified illustrations and extra large labels.

Your visualizations shall not be similar as advertising for cigarettes or cars, but their design should serve you at least a little bit as an example for the presentation-suited creation of your own visualizations.

Due to the reasons named above visualizations from books, journals and reports cannot be used directly in presentations in most cases, unless (which is an exception) they have already been created in a presentation-suited way!

Visualization tips
There is no "patent medicine" for the successful visualization of presentations, but a general direction to follow.

1. Slimming! Minimizing the message and contents of the image to the essential!
2. Makeup! Poster-like design and emphasis

The Slimming needs objectivity and creativity to find the essential and to formulate short and precise. This is also called didactic reduction.

The Makeup contains the following aspects:

- **Heading**: striking, meaningful, correct; contains central message; short, precise spoken language!!
- **Structure**: logical arrangement of 5–7 elements (words, items, figures)
- **Reading dynamic**: eye-catcher = starting point: if not in the upper left corner, then you should propose a reading sequence (direction, orientation) emphasized shape, color, size or "cloud"; end point: in the lower right corner, otherwise emphasize it.
- **Use of color**: wherever it makes sense, but well directed! Every color must be explainable! Keep color psychology in mind!
  (No riot of colors = sensory overload)
- **Font size**: varied between two sizes, but never too small!

▶ To test your visualization think about two aspects:

   Does the visualization work in DIN-A4-size from 1.8 m distance
   and after 3–5 s looking at it?
   (Put the visualization onto the floor and look at it when standing upright.)

After these rules and test criteria here is another important tip:

▶ The visualization must not make the presenting person superfluous!

What does that mean?

1. Literal reading of text bores your audience (they can do that themselves).
2. The labels in images should let room for the speaker's explanations!
3. Texts on slides should not be sentences, just keywords!
4. Personal, affected things should not appear on slides!

Regarding 1. and 2.: The speech shall be accompanied and emphasized by the visualization, but not replaced. The balance between information provided in images and in the speech must be carefully designed.

Regarding 3.: See the slides in the visualization example! The last slide is an exception, in the summary short mnemonics may be useful.

Regarding 4.: Words, sentences or formulas like "Good Morning, ...", "Any questions?" or "Many thanks for your attendance!" must be said and not shown. These are important means for your personal contact with the audience.

These hints in abridgment shall emphasize and complete the recommendations in Sect. 3.4. In addition, a specialty of the presentation visualization shall be addressed: the animation.

Animation

An animation (lat. sensitization, vitalization) we understand all that lives, i.e. all that moves in a visualization during the presentation. In the report there are only static images. The presentation is loosened, emphasized and improved at least by the image changes, but also quite frequently by the animation of parts or elements of an image (if this is not exaggerated).

Two good examples for animations:

1. List items appear one after another, blink or change their color when they are addressed.
2. Complex or time-dependent, discontinuous relations or processes, like the scheme of the production of a motor, the kinematic of a sewing machine or the material flow in a power plant can be displayed clearly (but costly), by emphasizing the current phase with a background color, a frame, strong colors or blinking.

   However: Avoid time-dependent animations, if you have to fight against stage fright! It is better, if you click each step on your own (wireless mouse)!

Poor animations

Constant movement on the projection wall, steady stimulus satiation by consecutively appearing texts, action and shapes and eventually additional background sounds like fizzling, humming or scraps of music, which let appear the presentation like a film and make the speaker superfluous.

Conclusion

The more possibilities of a modern presentation there are, the more sensitivity of the speaker is needed, to prevent losing the main effect of the presentation—the personal impression and the charisma of the person!

Two hints for the presentation with laptop and beamer:

1. Already today and even more in the future using a laptop and beamer is no attribute of quality any more. The fascination from the beginning gives way to customization. What adds up in the end are the technical and personal contents of your presentation again.
2. Presentation equipment may not operate properly.—Always take with you a set of overhead slides to important presentations!

Finally yet importantly, here is a small trick, which may help you in a computer-controlled beamer presentation: Let the presentation program display a timer with start at 00:00:00 in the lower left corner of the screen, and then you have an exact time control in sight, which helps you fight against the time. At the end of your presentation, you should answer the question, whether you want to save the show times with No.

Visualization example

Now our example "Burner-integrated suction when welding" will be visualized in eleven slides.

The basic layout of the slides in our visualization example shall contain the structure at the left margin. The current point of the presentation structure is emphasized with bold print. The emphasis can also be made with a different color like red. On the title slide (Fig. 5.7) and the structure slide (Fig. 5.8), the presentation structure at the left margin and the heading can be left out. The structuring lines can also be left out. Slide 3 (Fig. 5.9) shall emotionally involve the audience and visually support the introduction to the topic.

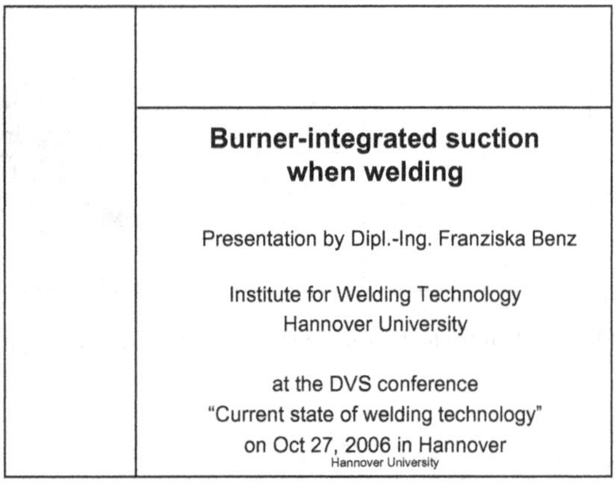

Fig. 5.7 Title slide (slide 1 of the example presentation)

Burner-integrated suction when welding

Presentation by Dipl.-Ing. Franziska Benz

Institute for Welding Technology
Hannover University

at the DVS conference
"Current state of welding technology"
on Oct 27, 2006 in Hannover
Hannover University

Structure

**Burner-integrated suction
when welding**

Franziska Benz, Hannover

Health vs. seam quality?

Suction of welding-pollutants
– burner-integrated
– optimized design
– verified quality

Summary and outlook

Hannover University

Fig. 5.8 Structure slide (slide 2 of the example presentation)

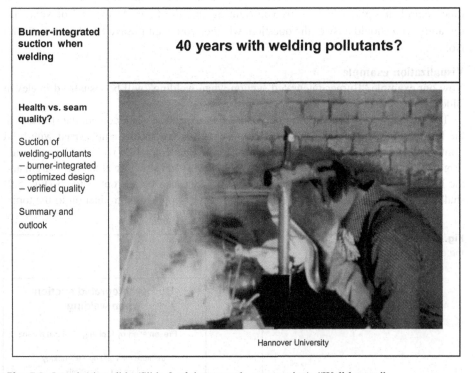

**Burner-integrated
suction when
welding**

40 years with welding pollutants?

**Health vs. seam
quality?**

Suction of
welding-pollutants
– burner-integrated
– optimized design
– verified quality

Summary and
outlook

Hannover University

Fig. 5.9 Introduction slide (Slide 3 of the example presentation), "Well-known"

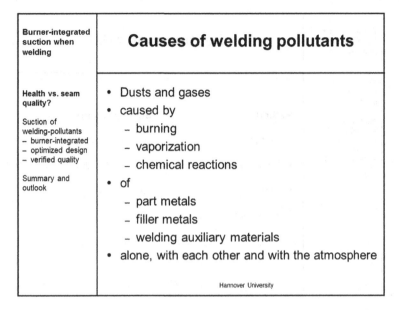

Fig. 5.10   Causes of welding pollutants (slide 4 of the example presentation), "Well-known"

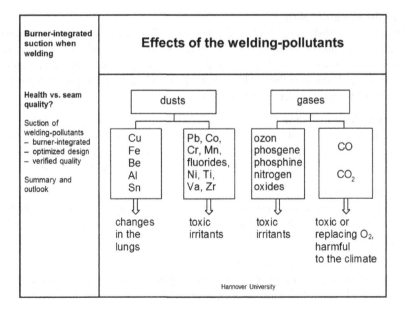

Fig. 5.11   Effects of the welding pollutants (slide 5 of the example presentation), "New"

Hints on the design of slide 4

Compact, but clearly many technical terms are listed in this slide (Fig.5.10) and explained in the presentation, i.e. the slides should not be totally self-explaining and should not contain complete sentences. If they were, what would you be needed for as the speaker?

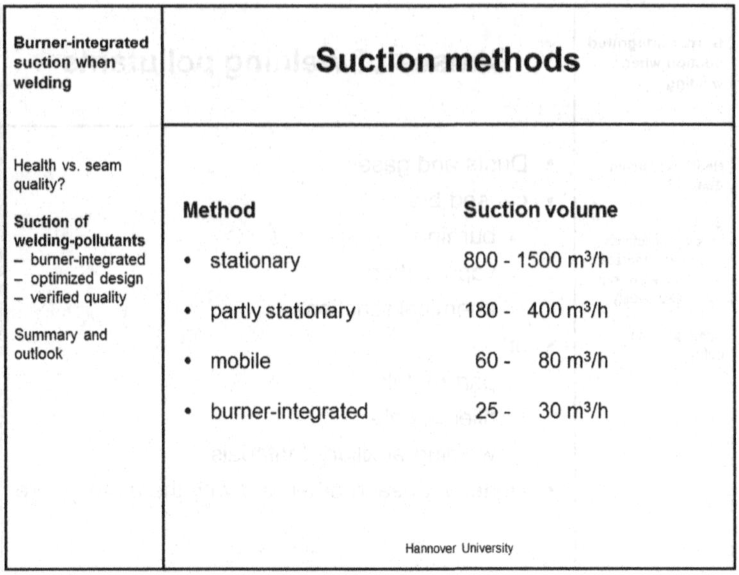

Burner-integrated suction when welding	Suction methods	
Health vs. seam quality? Suction of welding-pollutants – burner-integrated – optimized design – verified quality Summary and outlook	**Method**	**Suction volume**
	• stationary	800 - 1500 m³/h
	• partly stationary	180 - 400 m³/h
	• mobile	60 - 80 m³/h
	• burner-integrated	25 - 30 m³/h
	Hannover University	

Fig. 5.12 Suction methods (slide 6 of the example presentation), "New"

Hints on the design of slide 5
The design of the slide (Fig. 5.11) clear and tidy. The text boxes have the same size (different sizes would disturb the esthetic). When explaining the abbreviations of the dusts and gases, you can actively include the audience (those who know the abbreviations are happy; the others learn something). However, slide 5 ofthe example presentation can also be omitted or or kept as reserve, if time gets narrow.

Hints on the design of slide 6
On this slide (Fig. 5.12) you could define an animation for the two-column text, so that at first the items in the left column "Method" are displayed one after the other and then the four suction volume values in the right column are displayed (all together or one after the other).

Slide 7 (not shown here)
Slide 7 of the example presentation shows the conventional suction methods (hall, workplace, collector which must be positioned by the welder) in coarse sketches (Figs. 5.13 and 5.14).

Hint on the design of slides 8 and 9
The section drawings of the burner with the original nozzle (Fig 5.13) and optimized nozzle (Fig. 5.14) cannot be large and clear enough!

Slide 10 (not shown here)
Slide 10 of the example presentation shows the variation of the suction rings in sketches.

Hints on the design of slide 11
Slide 11 of the presentation (Fig. 5.15) and other slides, which are not shown here make up the theoretical, complex part for the experts ("Mazy"). Such images shall be more

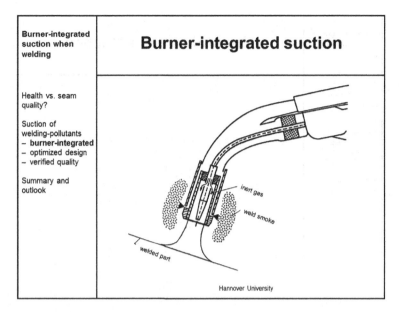

Fig. 5.13 Original suction nozzle (slide 8 of the example presentation), "New"

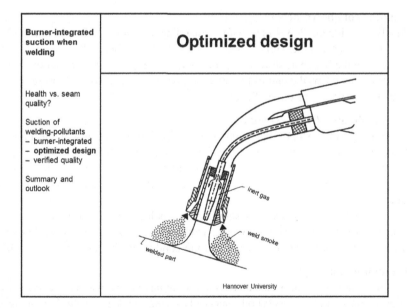

Fig. 5.14 Optimized suction nozzle (slide 9 of the example presentation), "New"

complex. The experts find their way through many figures, elements and symbols. A much higher information density is a part of the presentation tactic here: The experts study and learn the details and specialties while the majority is more or less overburdened for a not too long while and waits in awe.

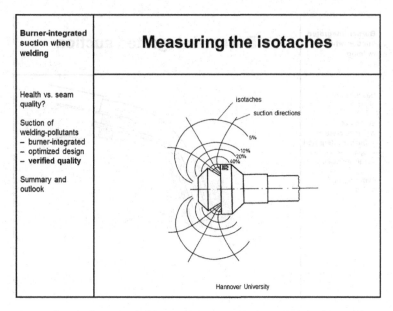

Fig. 5.15 Measuring the isotaches (slide 11 of the example presentation), "Mazy"

More slides (not shown here)
As already mentioned, 2–3 slides with demanding contents ("Mazy") and 1–2 slides showing the experiment design ("New") are not shown here. The second last slide (Fig. 5.16) with the bar chart showing the effectivity of the weld smoke suction in different welding positions forms the end of the main part of the presentation and brings new, clear messages to everybody in the audience.

Hint on the design of the last slide
In the summary (Fig. 5.17), all profound results or key statements of the presentation should be listed and spoken about without introducing essentially new aspects or information. With these points, the general understanding of the contents of the presentation shall be assured and a smooth transition to the further in-depth discussion shall be facilitated.

These examples of visualizations show you a small insight from the huge amount of possibilities. All decisions you make are a question of your personal preferences, the contents to be presented, your software and the conditions of your presentation task.

Presentation manuscript
This will be a short section, because you should not have a manuscript and speak out free (best impression). But not everybody succeeds in that, at least at the beginning of the career. Select the form of presentation manuscript you need as a guide for your thoughts. The following options are available:

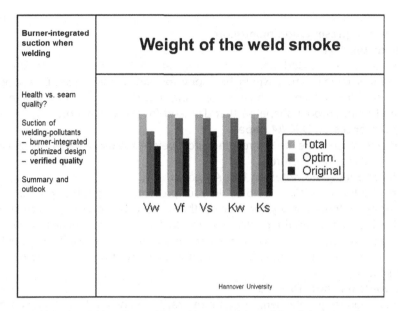

Fig. 5.16 Results of weighing the weld smoke (second last slide of the example presentation), "New"

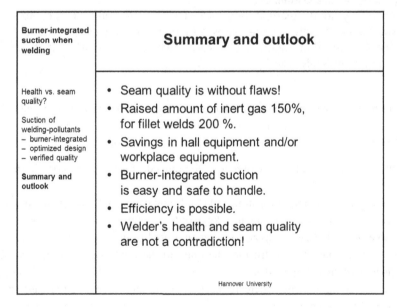

Fig. 5.17 Summary (last slide of the example presentation), "Nothing New!"

Forms of a presentation manuscript
Formulated text
This should be avoided, because it is a speech barrier. Exception 1: You read everything literally—that would be a pity for your presentation. Exception 2: You use a formulated text only for opening and closing your presentation, in case of being in poor shape on that day, and for literal citations.
Key sentences on DIN A4 paper
Mediocre, since there are still reading effort before speaking, paper rustling and constraints to free gestures.
Reduced slides and notes on DIN A5 paper
Open the function "View – Notes" in your presentation graphics program and write key sentences or keywords below the slides. Later you can print the notes and reduce them with the photocopier to DIN A5. These small pages can well be used as manuscript. This function is also very useful, if you want to speak in a foreign language. You can search relevant vocabulary at home and add them to the slides.
Keywords on cards DIN A 6
Rather good to good variant, because the cards can be easily exchanged when you create the manuscript and they are easy to handle after a little training. The handling of the cards does not cause much noise and implies only a few constraints to free gestures.
Keywords on the visualizations
Good to very good variant, if the keywords are written on folding cardboard or plastic bands directly on the slides and therefore not clearly visible to the audience. With that method, you can cause the impression that you are speaking completely "free".

From these options choose the one that is most sympathetic to you, try it out and make it perfect with growing training. The card method DIN A6, see Fig. 5.18, can be recommended for the beginning. Please keep the following hints and properties in mind:

- Your presentation can be combined, rearranged, made longer or shorter like a mosaique.
- For a new presentation you can use old and new cards (only the front sides!).
- On each card, you can note about 5 keywords or figures in as large (block) letters as possible (thick felt pen or large font size and bold print).
- The heading contains the structure item and in the upper right corner a consecutive number (written with pencil!).
- In the lower right corner, there is the running minute in which the card should be completely finished.
- The cards are punched for filing them in a ring binder.

The following color system has proven to be practical:

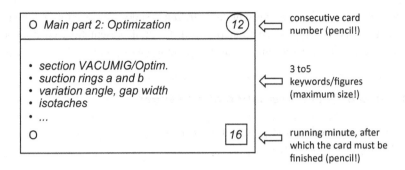

Fig. 5.18 Manuscript card

- Blue cards: "direction cards" for welcome, intermediate summaries (redundancy) and closing sentences
- Red cards: "must cards" for introductory words, main part and summary as well as all essential facts and figures
- Yellow cards: "should cards" for additions to the main part and summary
- Green cards: "can cards" as reserve material with additional information and details.

The system works as follows:

- Blue (direction), and red (must) cards are essential. They must be processed. Therefore it should be only as many as you can definitely use in the timeframe.
- Yellow (should) cards should be processed, if possible, but they can also be omitted (skipped), if you are running short of time.
- Green (can) cards are the reserve material, if you have more time than estimated (can happen as well) and if there are special questions.

The colors help to rearrange the structure during the presentation without losing the key message of your presentation (missed topic…).

Presentation handout or documentation
The transitions between manuscript and handout or documentation are floating. Depending on your information target you can hand out the following documents:

- the presentation,
- copied slides in original size or reduced (handout pages, 2 slides per page),
- note pages, if you did not only write a few keywords but more detailed information as addition to the slides,
- copies of various documents, which have something to do with your presentation topic, like copies of important figures or tables from your Technical Report, which are hard to project and/or
- an elaborated documentation.

Providing your audience with handout pages with 2 slides per page is the most common type of handout.

So far, we have created the presentation structure, slides, manuscript, handout and/or documentation. Now we proceed to step 6 with the trial presentation, changes and the necessary preparations, before the actual presentation takes place.

5.4.4 Step 6: Trial Presentation and Changes

Each presentation needs a trial presentation, if it is important and shall be successful. It should be organized as follows:

– All spoken parts including explanations of the figures and eventual showing films, videos or models/demonstration objects must be executed completely and realistic (no abbreviations, normal speed of speech and films).
– At least one person in the audience to …
– … write a protocol and take the time of all parts of speech and the total time.

It is best to execute the trial presentation in the later meeting room with its peculiarities, but all other rooms are also helpful. You should then evaluate the protocol honestly and objective and rework the presentation using the following checkpoints.

Evaluation of the trial presentation

– Where do I need to leave out what (if it took too long)?
– Where is it necessary to add something for better understanding?
– Which figures are too small, too overloaded, too complex?
– Which keywords, sentences, buzzwords or gags are useful,
– which are embarrassing or irritating? (Avoid irony!)
– Which contents do I have to add? Do I need reserve material?

Your audience does not need to be experts to judge a trial presentation. Sound intellect, interest and honesty are sufficient in most cases, to improve your presentation.

The right consequence of your trial presentation and its protocol is, that you are really willing to change everything what was not convincing. It is only then, that you have the feeling, that you did everything for the success of your presentation. In addition, the feeling of a good preparation makes you more self-confident (less stage fright), because you can show a good product. Even, if there is much to change—do it as good as you can.

Two and more trial presentations are quite usual e.g. for dissertation and habilitation presentations and audition lectures for the job as professors. This is because a radical cutback after the first trial presentation proves to be exaggerated in the second trial presentation and requires new additions etc.

5.4.5 Step 7: Updating the Presentation and Preparations in the Room

The crucial phase shortly before the presentation requires updating the presentation and the preparations in the meeting room.

Updating the presentation
Often the preparation of a presentation runs for quite a while. However, the contents of the presentation shall be really up-to-date, especially for the experts. This requires

- a last check of the contents of the presentation, especially the expert chapters,
- eventually a few looks into the latest editions of scientific journals and the newspaper and
- a talk with the professor, customer, boss or the conference organizers: there could be new conditions!

Preparations in the meeting room
On the day of your presentation you should be in the meeting room about two hours early, because there are still four tasks:

- the personal preparation,
- the technical preparation,
- the preparation of the meeting room, and
- the preparation of the contents.

Personal preparation
The personal preparation contains

- Be there early enough (traffic jams and missed trains are no excuse!),
- sleep long enough/be well-rested and fit/in good shape,
- fresh air in the lungs (take a walk),
- correct hairstyle and clothing (mirror!), and
- calm and self-confident charisma (try it).

Technical preparation
The technical preparation prior to the presentation consists of

- the image projection (test it, switch it on/off, adjust the image size, move the projector if necessary, adjust the focus, test the brightness and image sequence; take a spare bulb with you!),
- the position of the desk (good for speaking, correct height, not in the field of view; for right-handed persons it is best to have the desk on the right side of the field of view seen in the viewing direction of the audience; depositing rack for manuscript, slides, laptop, pencil, paper, pointer, stop watch),
- the microphone (position a fixed microphone, adjust it in speaking direction, watch out for the directional characteristic, eventually ask for it; fix a clip-on microphone and the radio transmitter to your clothing, switch it on, definitely test the volume!), and

– test to display models/presentation objects/film projectors (build them up, try them out and cover them again for a better dramaturgy).

Preparation of the meeting room
The preparation of the meeting room aims at a convenient atmosphere for the audience. They consist of the following steps:

– The brightness must be adjusted, so that the images look brilliant, but the audience can still read and write—eventually ask someone to help you.
– The air quality must be guaranteed by ventilation (otherwise, the audience might fall asleep).
– Adjust the temperature (or let it be done).
– Clean the whiteboard, or blackboard, remove used paper from flipcharts.
– Near the desk, remove everything that might detract you.
– Care for cleanliness and tidiness and
– Remove sources of noise like a loud air conditioning, noise in a neighbored room, close windows against noise from the street etc. (or let it be done).

You might ask: "What are janitors good for?!" That's ok, ask him for help, but during your presentation you are responsible!

Preparation of the contents
The preparation of the contents prior to the presentation consists of

– distributing the structure or documentation and
– eventually showing an introductory slide (like slide 3 of the example presentation) or a slide signaling a break or a welcome slide.

5.4.6 Step 8: Lecture, Presentation

This step means actually presenting your slides and keeping all rules and hints in mind. In Sect. 5.5 it will be further divided into the phases contact preparations and contacting the audience, creating a relationship with the audience, appropriate pointing, and dealing with intermediate questions.

5.5 Giving the Presentation

Finally, there is the start.—The nearly unbearable tension can relax.—But how can you reach that? How do you warm up? How do you survive the first critical five minutes? Read and try out the following experiences, rules and tips?

5.5.1 Contact Preparations and Contacting the Audience

Your presentation starts with contact preparations and contacting the audience.

Contact preparations
The speaker goes to the desk well dressed, well rested, firm, and self-confident. She starts the stopwatch on the desk (is often forgotten!).

Do not use a wristwatch with real time display, because you would often have to compute the speaking time, which disturbs you.

The speaker looks around to the audience once, smiling friendly and respectful, looks at the persons in the first row and says:

Contacting the audience
"Doctor Gärtner (president of the German association for welding technology), professor Hauser (chairman of the institute), dear ladies and gentlemen, I can also see the sun shining outdoor! All the more I am happy for your turning up in such large number to listen to my presentation!"

These steps "contact preparations" and "contacting the audience" aim at a positive optical impression and thus acceptance of the person and willingness to listen to the contents of the presentation.

These two phases should not be underestimated: Those who appear chaotic, arrogant or very shy gain the first prejudice, disadvantage and drawback.

When greeting your audience it is important to keep the correct order of prominent people and the rest of the audience:

- The "main person" comes first (ladies before gentlemen, the district administrator before the mayor etc.).
- Maximum 5–7 people should be individually greeted, otherwise name groups: "Dear assemblymen, dear professors".
- "Main person" or "prominent" are also persons, which stand out somehow: one lady amongst gentlemen, one man amongst women.
- Politicians should always be mentioned with their names (they want to be recognized!)
- Do not name anybody, who is not present! (Awkward, drawback for bad preparation, see Sect. 5.4.5).
- Name titles, but do not exaggerate.

In doubt ask the conference chairman or the secretary to make sure that you greet the prominents with their correct titles and in the correct order—they will be happy to support you. In a negative case the "important" people, who are also often decision makers, are right away from the beginning offended in their vanity and start to revolt in their mind against your presentation. Moreover, the audience will take your savoir vivre as a hint for the potential of your company. Do not start any experiments in this phase!

The presentation starts now with an introduction of the speaker, the presentation target and how the topic shall be approached by means of the structure.

Opening

"My name is Franziska Benz, I am research assistant at the institute for welding technology of Hannover University since one year. My presentation target is to show you the latest findings of burner-integrated suction when welding. I want to proceed according to the following structure:"

(After the title slide and the structure slide have been shown, slide 3 appears now, and it is explained without literal reading.)

This businesslike opening is always all right, if you do not have a more thrilling idea. If you want to start with a gag ("wow effect"), plan and test it carefully in front of honest, critical test persons. If you introduce yourself, you never name your own title, unless you are in need of it ...

Before the presentation, you can literally formulate the greetings and read them literally in case of "start problems". You should always write down the names and titles of the important persons to greet and literally read them from your note.

5.5.2 Creating a Relationship with the Audience

In the human communication, also when presenting the most boring technical contents, it is inevitable to create a positive relationship between speaker and audience. If there is nothing "going on" between both parties, the presentation does not "come across" well!

This experience is often disregarded by technicians, because they are so convinced of the quality of their subject and qualification, that they do not care enough for "these manners". However, "these manners" are more than a container for the technical contents. The manners, the art of human interaction in the widest sense, are the key to introduce technology to people and to influence the decision makers in a positive way. Therefore, engineers should know a little bit of psychology. The more important teams become in our professional life, the more important are key competences like these.

How can you build up such a relationship at the beginning of your presentation? There is no patent medicine, but a recommendation: Friendliness, open-mindedness and interest for the audience create sympathy and in response open-mindedness and acceptance for your person and your presentation contents in the audience. Here are some means and methods for creating a positive relationship as examples:

- a successful, well-balanced greeting,
- referring to the common situation ("nice room", "yesterday night we visited ... together" etc.),
- integration of contents from the previous presentation, praise of the previous presentation (if appropriate) and advertising for the next presentation,
- friendly, positive, human introductory sentences,
- offer to ask intermediate questions, organization of breaks, offer to open the discussion,
- distribution of the structure or a first documentation (1st present),
- announcing/promising a more bulky documentation or a little surprise at the end of the presentation or

– anything else positive that comes to your mind …

In any person, also in yourself, is a child that wants to be hugged—then they are much more open for the message of your presentation.

Thanks and introduction
Our speaker thanks the head of the institute, her supervising research assistant and some laboratory engineers for the good professional and personal supervision. After these thanks the speaker presents the complete introduction by showing slide 3 and underlining the necessity of health and safety at work when welding. To create concern in the audience she asks the rhetoric question who would like to have his father, brother or sun suffer in such smoke for a whole professional life.

5.5.3 Appropriate Pointing

When explaining the figures, you should keep as much eye contact with your audience as possible, i.e. you either show with a pointer (plastic hand, flat ruler, tapered, not rolling pencil) on an overhead projector or with the mouse cursor on the display of the laptop. (Speakers who use a beamer cannot avoid to point with a cursor.) It was the idea of the overhead projector to point while looking to the audience! Telescopic pointer and laser pointer should stay at home to avoid speaking "with the projection wall".

If you point on an overhead projector with a pencil, lay it down onto the slide, let it loose and move it to point something else, the audience can see the pointer direction longer, more sharp-edged and better. If you keep the pencil in your hand and tip on the slide, the audience can recognize with pleasure the trembling of your hand (strongly magnified) and can imagine your stage freight very well. The latter is especially true for the often wildly dancing red point of the laser pointer.

Now the first intermediate question is asked. What are you doing?

5.5.4 Dealing with Intermediate Questions

Generally speaking a modern audience does not want to be silenced. Intermediate questions belong to the presentation time and cannot be forbidden rigorously (that seems unsecure, inflexible and dictatorial; hence, you have to plan some time for that!). There are two big categories of questions: real questions and unreal questions.

Real questions can refer to the organization, bad understanding, problems or just be difficult.

– You should answer questions referring to the organization (e.g. switching on the room light) and questions due to bad understanding (e.g. a word or figure was not understood) at once, friendly and short.
– You should check problem questions, which require a longer answer, whether you want to answer them at once or after the presentation. Then you should gain time by

repeating the question for the whole audience. Then you should answer the question short and precise or ask, whether it is acceptable to answer the question after the presentation (note the question!).
– Difficult questions, which you cannot answer should either not be answered or you just announce an estimation or assumption ("Please do not nail me down to it!"). Pass the question on to the audience—often someone knows it.

Unreal questions are no questions, but opinions, self-expressions, objections or pure distortions, e.g. by competitors. Countermeasures:

– recognize that in time (a matter of training), react friendly, but tight,
– depending on the situation "answer", appease or refuse that,
– do not allow yourself to be provoked,
– eventually let the audience vote and in any case ...
– stay able to pull the strings!

Finish
Our speaker shows the last slide, explains it and says while opening the last slide with her communication data: "Dear ladies and gentlemen, I am happy, that you have listened so curious. This was my first presentation, in the beginning, I was very excited; but now I feel fine. I thank you for your attentiveness and would like to answer all your questions now. Many thanks!"—Applause—now the speaker hands out her detailed documentation.
    Always set a clear finish and do not let your presentation "melt away"! The audience needs a clear signal for its applause, and this signal should come from you.

5.6 Review and Analysis of the Presentation

Is it possible to review such a complex and diverse activity like a presentation? What is more important? The contents or the show, the rhetoric? The opinions of experts will vary here. Nevertheless, we want to try to evaluate and review a presentation as appropriate, selective and objective as possible, even if it is only for educational purposes. The basic idea of the following evaluation scheme is the balanced weighting of all elements of a technical or scientific presentation, Table 5.8.

Table 5.8 Evaluation scheme for a technical presentation with weighing the presentation elements

Presentation elements	Weight
Performance (including rhetoric)	30% weight
Contents (amount and level)	30% weight
Organization	20% weight
Impression (subjective)	20% weight
	100% result

These figures have been selected from experience: they have proven to work well. The weights express, that the rhetoric or the contents alone do not make up the total quality of a presentation. Similar to the analysis of a human person the sum of all properties matters. This does not make the task easy, because you have to care for so many things at the same time. On the other hand, this is a chance for every speaker, to emphasize his/her strengths and to compensate weaknesses.

The following scheme would probably look different for other disciplines like humanities. At the same time, it is a property list or checklist for the preparation of a presentation. The 25 criteria will now be shortly defined.

Checklist for the evaluation of the final presentation (criteria)
Performance

– Atmosphere: interpersonal and meeting room conditions
– Introduction: contacting the audience, introduction, eventually "wow effect"
– Rhetoric:

  – Speech flow: velocity, variations, breaks
  – Volume: too calm, nice, too loud
  – Understandability: distinctiveness of the voice, accentuation
  – Mimic: facial expression
  – Gestures: Usage of hand, head and body motions
  – Standing: firmness, posture
  – Eye contact: eye contact and leading the audience
  – Charisma: impression of personality, engagement and persuasive power

– Tension: tension flow, dramaturgy
– Flexibility: reaction after questions, slip-ups, distractions
– Finish: recognizability, harmony, conclusion

Contents:

– Amount: too much, well, too few (no reserve material)
– Level: too high, well, too low (bad mixture)

Organization:

– Preparation: material selection, effort, planning
– Structure: structure, logical design
– Transparency: clarity, straightforwardness, recognizability of the structure items
– Visualization: visual display of important facts
– Usage of Media: selection and usage of media (blackboard or whiteboard, flipchart, overhead projector, laptop and beamer, models)
– Redundancy: repetitions and intermediate summaries
– Timing: duration of the presentation parts, keeping the lower/upper time limits

Impression (subjective):

- Learning success: "My gain of knowledge and understanding: ..."
- Identification: "I have adopted the following facts, opinions and arguments: ..."
- Motivation: "My desire to deal with this topic, learn more and eventually plead for it: ..."

Assessment

Using the above criteria, you can calculate as follows.

(a) For the fulfillment of the criteria above you can give the following grades:

- very good, very often, accurate: grade 1
- good, quite often, rather accurate: grade 2
- average, mean, a little inaccurate: grade 3
- little, seldom, low, rather inaccurate: grade 4
- not acceptable, not visible, very inaccurate: grade 5

(b) Form intermediate sums for each group of criteria (1–4).
(c) Finally weigh the intermediate sums to get the result.

- intermediate sum 1 (performance) /13 × 0,3 = ...
- intermediate sum 2 (contents) /2 × 0,3 = ...
- intermediate sum 3 (organization) /7 × 0,2 = ...
- intermediate sum 4 (impression) /3 × 0,2 = ...
- result = sum total of the four weighed intermediate sums

Review

This evaluation scheme delivered the following results or weaknesses in more than 800 presentations:

- There should always be a handout with the structure lying in front of everybody in the audience!
- A precondition for a serious, successful presentation is good clothing.
- At the beginning, it is important to greet all celebrities and others in the audience correctly and to build up curiosity and tension by announcements, promises or questions. These do not appear by themselves!
- Above every image at the wall, there should be the appertaining chapter heading or a heading that can be easily assigned to the current structure item to provide the audience with calming transparency.
- The rhetoric is mostly good, but sometimes the speaker literally reads everything or at least too much (write only keywords and figures into the manuscript!). Another weakness is a lack of mimic and gestures (due to stage freight, too much concentration

on the topic and a not very relaxed attitude). These are typical errors of technicians when presenting.

- Often the contents does not contain the three levels—Well-known, Mazy and New. Then the contents is either too superficial (it contains only few expert knowledge, i.e. no details, specialties, insider experiences) or it assumes too much as "a matter of course" (also it often contains too many abbreviations and too much technical jargon).
- The presentation often contains too much theory at the beginning and too few examples, descriptions, stories (better is to first bring practice and descriptive stuff that create curiosity for the theory!).
- Nearly always, the redundancy (summaries and short repetitions between the chapters) runs short—it must explicitly be integrated into the manuscript!
- Often there is no or too little reserve material prepared, with which a too short presentation can be inconspicuously expanded to the right amount of time.
- The organization is often too intransparent and has a lack of redundancy and time problems.

Try to keep these very important points in mind and to avoid any weaknesses. But: One step after the other! It needs much training and a few bitter experiences from own presentations, until you are perfect. So, after your presentation you should write down all strengths and weaknesses of your presentation immediately. If you regard these items for your next presentation, you will steadily become better and have more and more success and fun when presenting.

5.7 The Short Statement

Quite often in our present business life, you have to explain an issue to an audience relatively short and occasionally extemporized. This is the opportunity for a short statement of 3, 5, or 10 min duration. Its quality regarding form and contents does not differ very much from the classic prepared 20 min presentation, as it has been described before, but there are a few details different.

5.7.1 Trigger

Often you get the task to give a statement spontaneously: "Mr. Miller, please explain us, which tasks your team performs!"

This order by the chairperson of a meeting can occur at any time more or less surprising, be adumbrated from the agenda or cleverly foreseen. Another example: "Dr. Kluge, please inform us shortly about the present situation."

Now you as the speaker have to decide depending on your audience whether you give a 3 min or 10 min statement.

5.7.2 Requirements

For the presentation of a statement you have to distinguish two cases:

then Shift+Enter Case 1: All participants are familiar customers and colleagues—the audience know you.

then Shift+Enter Case 2: At least one person from the audience does not know you.

In the seldom case 1 (everybody knows you), the approach is simpler, because you can leave out the introduction of your person and tasks. The speaking time is 3–5 min.

In the frequent case 2, you are not known by everybody and should execute the contact preparations and contacting the audience even in the short statement, to give more weight to the contents of your remarks. The importance of the message increases, if all members of the audience rank the messenger and his credibility in a positive way.

5.7.3 Example Statement for a Familiar Audience

Now you see an example statement for case 1. Mr. Miller explains the following about the work of his team:

I want to speak about two points and make a concrete suggestion.

Point 1 is: The design work has developed well, most requirements (about 90%) of the specification sheet are fulfilled, partly even over-performed, which is true for strength and safety against rupture.

Point 2 is: Design and price are not yet definitely settled. The machine seems to be unpractical and homespun. The ergonomics must be improved. This is also true for the look-and-feel. Our design line should be carved out better.

So far, our price is quite high as compared with our competitors. Therefore, it needs a recalculation.

My suggestion: We design the housing more elegant and less expensive with a different plastic material. This implies concessions to the safety against rupture, but the selling price can be reduced by 20%.

We will need about 8 weeks for these changes.

Are there questions? No?! Thank you very much!

The essential characteristics if this 3 min speach are:

- It is mainly fact-oriented and unemotional.
- No greeting, no introduction of the own person and the tasks, because they are known to everybody.
- Clearly structured into three parts (announcement/introduction, main part, conclusion/questions).
- Each part is clearly visible. This creates transparency.
- The announcement of the suggestion creates expectation and attention.
- Statements and explanations are conclusive.

5.7.4   Example Statement for an Unfamiliar Audience

Here an example statement for case 2 will follow. Dr. Kluge informs about the present situation of her project:

> Director, my honored guests,
>
> my name is Kluge, my department HG 2 designs top class household appliances, which last year brought our company to market leadership. I am glad about your interest. If you have questions, you may interrupt me.
>
> First I show you the current state of the design, second the work steps to be done and third my planning up to selling.
>
> On the first slide, you see …
>
> (as in case 1)
>
> Finally, I want to thank you for your interest; I am grateful for your suggestions and hints. I think most of the proposals can be realized with the help of my young, flexible team, which I want to praise here.
>
> If you have more questions, my team members are at your service. You can reach them using the following contact data: …
>
> Thank you for your interest and on continuously well cooperation!

The characteristics of this 10-min speech are:

- Strong involvement of the personality by introduction, description of the tasks, contacting the audience, positive charisma.
- Broad contents (clear structure, high transparency, detailed statements, and sound explanations).
- Personal conclusion, audience gets "desire for more" from the offer to continuously work together well.

5.7.5   Problems of Short Statements

In conjunction with short statements, the following five problems occur repeatedly:

- Improvisation is often required. To pass the situation well, you can only preplan all possible situations and hold slides or manuscript cards ready.
- In most cases, it is not possible to create a tension flow.
- Quite often, the audience is impatient. The same holds true for the bosses.
- Often there are no devices. You have to speak offhand. To prepare yourself for that, you can only practice improvisation, e.g. a speech about the advantages of a torch or possible uses of paper clips.
- The timeframe is often unknown. If possible, clarify that in advance, e.g. instantly after you got the invitation to take part in a larger meeting. Otherwise, ask for it at the beginning of your short statement.

5.7.6 Tactical Measures

To be prepared for all eventualities, the following measures have been proven to be helpful:

- Hold short structures on note cards and manuscript cards in readiness.
- Plan to speak 1 min about one card, no more.
- The time partition should reserve about 20% for greeting and introduction, 70% for the contents, and 10% for the conclusion.
- Practice to give short statements with beamer or magnetic board or magnetic board or magnetic strip, also sitting, or better standing at the table with cards in your hand or lying on the table.
- Make clear transitions between the items of your speech.
- Prefer to have much eye contact with your audience, read little from your cards.
- Formulate your sentences freely.
- Observe and practice the rhetoric tips in the short statement as well.

5.8   57 Rhetoric Tips from A to Z

On popular demand of our readers, here are a few rhetoric tips from A to Z, which are partly derived from the three books listed at the end of this chapter.

ABLR5 does not mean something to everybody—explain all **abbreviations**, otherwise you will earn questions that cost you time and disturb you.

Audience: the addressees of your presentation, but also participants and partners, who want to be treated like that and like humans, not like scanners. Otherwise, the audience goes on strike and you talk against a wall.

Accentuation (the right one) is inevitable for a vivid presentation style and for understanding, especially for foreign words and proper names (when in doubt, look them up or ask someone).

Arrogance is out of place even for a total expert. It makes you dislikable and disturbs the acceptance of the audience for your message.

Body language via gestures and mimic turns a presentation into an event and experience for the audience that continues to have an effect.

Breathing should not be done like a singer, but imperceptibly; that means as short and often as possible or "quick, silent and without effort".

Chance: Every successful presentation provides you with it—a good critique, a credit, an order, funding or a career step may follow.

Charisma of the speaker: It is based on competency in the field, engagement, sovereignty and an open and friendly communication, see also personality.

Clothing is the first optical impression; it shows your engagement for the audience; it shall be adequate to the situation, the contents and your person, when in doubt preferably traditional and better than usual. No experiments!

Contact to the audience is the main advantage of the speaker in contrast with a book or video. Design it with care by greeting, eye contact, questions and personal remarks. To do that, do never use images! ("Many thanks for your attention!" on the last slide is contra-productive but it appears often in presentations.).

Controlling himself and the audience is hard for the speaker, but inevitable: time, attention of the audience as well as the quality of speech and images must be constantly verified.

Demagogy may never be the presentation or speech target, so do not present false or unfair contents!

Dialect is human and you can hear it nearly everywhere; it should be neither suppressed nor exaggerated; it shall not disturb the understanding of the audience, not seem ridiculous and not part. Ideal is a "thought high-level language".

Eye contact is the first bridge to the audience, creating contact, respect and trust; these are the prerequisites to create acceptance in the audience.

Feet are visible in most cases and part of your body language, use them in a natural and inconspicuous way.

Filler words and phonemes are dispensable and do not show discipline of thinking! Instead of beginning a sentence with "Er(r)", "Well, …", "And…" and "Actually…" you should better make a pause.

Foreign words should always be explained or translated (along the way with ease) before someone can knit one's brows, that would just cost seconds and avoids questions. Using foreign words is no guarantee for scientifically correct information.

Formulation is the most important means to achieve understanding. Form clear, short sentences in personal colloquial language and avoid a too scientific, too formal, or overblown speech style.

Gestures are the support of your presentation contents by means of body language. Holding your hands in the height of the bellybutton is the best and easiest initial position for optically emphasizing important (not all) items.

Greeting is the first verbal contact with the audience. It shall be planned carefully (celebrities…) and easily and winningly build the first bridge, but it should never be read from a manuscript, possibly even without eye contact.

Hands are a problem only at the beginning. They should be always visible and as often as possible without holding things (cards, pointer, mouse) to be able to make gestures.

Humor is the spice of your speech; it should never miss completely (also humorous self-criticism), always be well tested and never be embarrassing or racist.

Impression is the image you create in your audience. To achieve a good one you may exaggerate from time to time, behave different or act a little.

Influencing is one aim of good rhetoric. The audience shall be motivated to understand, to positively decide (funding, buying, prolongation of the project,...), to personal trust and to appreciate the speaker.

Inhibitions are minimized by excellent preparation (trial presentations) and best physical fitness (sleep, clothing, punctuality).

Intelligence is not measured by your scientific vocabulary, but by your flexibility, quick-wittedness and art of human interaction.

Intermediate questions should be answered friendly, completely and quickly to cost less time.

Key and foreign words: They should be spoken out very, very clearly and embraced in speech pauses.

Listening to yourself is the recipe for unsuspiciously checking your speech, the clarity, volume and speed: From time to time you should listen to yourself while you are speaking to minimize problems.

Mimic is the language of your facial features; it shall be vivid and natural, i.e. change it from friendly to neutral or serious from time to time, depending on the presentation contents.

Modesty is the opposite of arrogance; but too much of it can be interpreted as shyness (uncertainty). Ideally you are modestly self-confident ("I know much, but not everything.").

Pauses are the best of work. Also in a presentation, moments of silence are appropriate: pauses of thinking, pauses to emphasize something, pauses to let something continue to have an effect, but do not make too long pauses in moments of uncertainty or blackout (and never use filler phonemes, see also filler words and phonemes!).

Personality is your main capital for the success of a good technical presentation. It is not only important, what you know, but also who you are or what the impression of your audience is, see also charisma.

Persuading by speech and supporting gestures and mimic is the result of a good presentation—people must be able to believe, what you are saying.

Pitch of the voice and speech intensity should vary during your presentation to avoid monotony. A comma in the sentence is expressed with higher pitch and a period always with accentuated and decisively lower pitch. Especially a good closing sentence ends in totally low pitch.

Point is the peak of humor; it should be really witty, appropriate and tested, otherwise it is likely to be a disadvantage.

Pointing should never be done against the wall, but while keeping control over the audience with a tapered ruler on the projector or with the mouse on the screen of the laptop!—Point calmly and longer, because not everybody in the audience looks to the front in the same second!

Posture is the optical means to signal security, decisiveness and modesty, i.e. stand upright and tense, but still vivid and always with your front towards the audience.

Questions are the icing on the cake, but they can also freeze your presentation (see Sect. 5.5.4). Take it as a racy challenge and prepare necessary answers in advance.

Repetition of important contents, figures and facts supports understanding and remembering.

Review is clever to learn from every failure and success; this is the only way to improvement and routine.

Sayings can be integrated very well, if they fit to the topic. Especially you can tell (or show) the first half of a saying at the beginning of the presentation and the second half at the end to create tension.

Sentence structure should be simple and easy to understand, should not contain nested sentences, see formulation.

Slip-ups can happen, anticipate them! (slip of the tongue, malfunction of the projection technology, mismatch of images…). Do not exaggerate small slip-ups, but address bigger ones and beg for apology.

Smiling should spice the whole presentation where it fits without appearing too sweet. Technical and serious facts come across better with a friendly face. When there are disturbances, you should always stay friendly (=sympathetic), even if you are very, very upset.

Speech style: It should be oriented towards the middle of the society—there are the decision-makers!

Spoken language does not mean small talk or chitchat, not the speech style of a user's manual, a law or a government declaration, but a selected, generally understandable everyday speech.

Stage-freight, see inhibitions!

Standing shows the condition of the speaker. You should stand in a natural position, neither stark and stiff nor changing all the time. Then it disturbs the least from your presentation.

Studying all speeches and presentations in the public (TV talk masters, politicians, experts…) helps to improve yourself. In addition, you should always study your audience for their reactions.

Tension within a presentation—if not planned into the contents—can also be created with your voice, mimic and body language.

Time pressure is the strongest enemy of the speaker. The only precaution are a stop watch and a lot of training to defeat this enemy.

Training is more important than eloquence; frequent presentations and speeches with honest feedback of an audience create experience and routine.

Trust creation is an important target of your presentation—trust into the contents, your person and your company.

Velocity of the speech should be adopted to the contents and its importance; when in doubt speak slower than usual and check by listening to yourself!

Volume (of your speech) depends on the size of the meeting room and the number of people in the audience; it should preferably be louder, but vary from time to time.

5.9   References in This Chapter

Herrman, P.: Reden wie ein Profi. München, Orbis, 1992

Jung, H.: Versammlung und Diskussion. München, Goldmann, 1980

Brehler, R.: Modernes Redetraining. München, Falken, 1995

DIN 19045-3:1998-12 Projektion für Steh- und Laufbild – Teil 3: Mindestmaße für kleinste Bildelemente, Linienbreiten, Schrift- und Bildzeichengrößen in Originalvorlagen für die Projektion. (Projection for still and moving images – Part 3: Minimum measures for smallest image elements, line thicknesses, character and pictogram sizes in master copies for projection.)

Summary

6

Now all details, rules and working procedures relevant for writing and presenting Technical Reports have been introduced in detail. Our network plan for creating Technical Reports has been processed from accepting and analyzing the task to distributing the final report. It has also been shown in detail to present a Technical Report to an audience or hold a lecture about its contents, Fig. 6.1.

During all work steps the creator of the Technical Report must always ask at first, whether there are rules issued by the customer or already existing within the own institution, how Technical Reports must be written and designed. When using this book, please keep in mind: Already existing rules (standards of the department or professor or company or customer) must be followed prior to the tips and rules provided in this book.

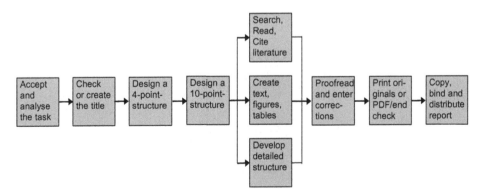

Fig. 6.1 Network plan for creating a technical report: All working items are finished

© Springer-Verlag GmbH Germany, part of Springer Nature 2019   227
H. Hering, *How to Write Technical Reports*,
https://doi.org/10.1007/978-3-662-58107-0_6

If such institutional standards do not exist or do not cover all details, you should use the hints and suggestions in this book. The consequent application of the information and working procedures described in this book will probably improve the quality of your future Technical Reports as compared with the quality of your previous ones.

We wish you that your future Technical Reports and their presentation will contribute to your personal success!

Glossary—Terms of Printing Technology

In the following we will shortly explain important terms in the field of printing technology, which might help you to design your Technical Reports and when you are in contact with copy-chops, computer stores, printers or journal and book publishers.

Acrobat Reader is a free-of-charge viewer from the company Adobe to read PDF files, which you can download at www.adobe.com/de/products/acrobat/readstep2.html
The **ANSI** (American National Standards Institute) deals with standardization and has e. g. developed the → ASCII Code.The **ASCII** Code (American Standard Code for Information Interchange) is a standard to store letters, figures and symbols as decimal number in Byte (in txt files). It can only code the European languages.

BW is sometimes used as an abbreviation for black and white
Bookmarks are used in the internet browser (favorite internet addresses) and in PDF files (navigation through the document on the left side).
Bullets mark the different items in unstructured lists. People either use the common symbols •, •, -, –, — etc. or more "pictorial" symbols from fonts like Symbol, Webdings, Wingdings, Zapf Dingbats etc

Caps, Capital letters → **majuscles CMYK** stands for the colors cyan (turquoise), magenta (pink), yellow and black. Printers need all color information based on this color system for 4-color-printing.
During **compression** the file size is reduced by a packing program. Moreover, many single files can be combined to one easy-to-handle archive file. In this process a file with the extension *.zip is created. Due to the compression more data can be stored on the storage device. Sending e-mails with a compressed file in the attachment is much faster than without compression.
Consistency in Technical Reports means, that equal tasks regarding spelling, punctuation and typography are performed always in the same way throughout the whole report.

© Springer-Verlag GmbH Germany, part of Springer Nature 2019
H. Hering, *How to Write Technical Reports*,
https://doi.org/10.1007/978-3-662-58107-0

Correction symbols are used during proof-reading and are standardized in ISO 5776. **cpi** (characters per inch) . 10 and 12 cpi are common spacings for fonts with fixed spacing (typewriter or fonts like Courier, Letter Gothic etc.).

dpi (dots per inch) 300 and 600 dpi are typical resolutions for laser and ink jet printers. 300 dpi is the minimum resolution of images, if they shall be printed in a journal or book.

Digital Object Identifier **DOI** for electronic documents

In the internet digital documents (files) or objects can be found under the address, where they are put on an internet server (URL/URI), but internet addresses are changed from time to time. Therefore the DOI (digital object identification) system was founded by the International DOI Foundation (IDF) to identify the objects themselves. The objects get a number and a server looks in his data base, where the objects are currently available. Example: The object with https://doi.org/10.1007/s003390201377 is found, if you go to the DOI Resolver and enter the DOI number into the search window of the server or you enter directly the URI https://doi.org/10.1007/s003390201377 into your browser (i.e. http://dx.doi.org/ and DOI number). You will then be redirected to the server, where the object is available or get a list of links.

DTP (Desktop Publishing) is preparing documents at your desk ready to be printed by means of suited DTP programs which can mix text and graphics. Creating two and more column pages and the positioning of figures can be controlled more precisely than with a word-processing program and for later printing → CMYK color separations can be created. Quark Express and PageMaker are well-known page-oriented DTP programs, with FrameMaker you can create even very thick documents or books very efficiently.

DTV (Didactic Typographic Visualization) according to REICHERT is visualizing with „text images" (pictorial and tabular re-arrangement of text). Here unstructured lists, boxes, arrows and similar typographic means are used. DTV also makes clear the logical and evtl. hierarchical dependence of text blocks by means of targeted usage of lines, which mostly run vertical and horizontal only. DTV fills the gap between conventional continuous text and graphic visualization and allows easy creation of the text images, interested reading and spontaneous understanding of the relevant facts.

A **fixed spacing** font is a font with a fixed character spacing (Courier etc.). In a fixed spacing font the distance from the middle of one letter to the middle of the next letter is constant. Therefore there is some white space on the right and left side of narrow characters (opposite: → proportional font).

Font attributes define the typesetting of the characters and words in a certain font. Font attributes are for example, **bold**, *italic*, underlining: <u>simple</u>, <u>double</u> and dotted, but also crossed out, ^{superscript}, _{subscript}, expanded, SMALL CAPS and CAPITAL LETTERS.

The word glossary is derived from the Greek γλωττα = tongue, language. It is the international name for an alphabetically sorted list of technical terms together with definitions of these terms.

Half tone image is an image with smooth transitions of the gray or color values.

The **header** of a table is the top row (line). It contains the generic terms of the entries in the appertaining column. Therefore, it is often accentuated by typographic means (e. g. with a double line or with a gray background).

A **hyperlink** is connection, a reference or a goto command. If you click on the hyperlink, you go to another position within the same file or a program is started and the file or internet address named in the hyperlink is displayed.

Hyphen → hyphenation proposal (soft hyphen, SHY) or → non-breaking hyphen, NHY

A soft hyphen (SHY) is a **hyphenation proposal**, which is entered in the middle of a word to prevent wrong automatic hyphenation or to avoid too large distances between the words. If due to text insertions or deletions the soft hyphen moves to the middle of the line, it is invisible (in the printout). If you enter a normal hyphen and it moves to the middle of the line, it is visible and needs to be deleted explicitly via keyboard, what is often overseen or forgotten.

An icon is a small pictorial symbol on the desktop or in a computer program. If you click on the icon, the computer runs a function. That is more convenient than using the menus.

An **inch** is a unit of length. 1 inch = 25.4 mm = 2.54 cm. The resolution of images, printers, copiers and scanners is specified in → **dpi** (dots per inch).

The word **index** is Latin and means forefinger, pointer, overview, title and table of contents. It is the international name for the list of keywords with page numbers to quickly find the relevant text passages.

The **introductory column** of a table is the first column from the left. It contains the generic terms of the entries in the rows. Therefore, it is often accentuated by typographic means (e. g. with a double line or with a gray background).

International Standard Book Number **ISBN** for books (monographs)

International Standard Serial Number **ISSN** (for journals and serial publications)

The term layout summarizes all measures to influence the appearance of information on the paper. This includes the document or page layout resp. (for example defining the page margins, usage of a page header) and the definition of paragraph and character formats: selection of the font type and size for document part headings, text, figure subheadings, table headings, indentions, accentuations (italic, bold, underlined), usage of bullets in unstructured lists as well as the definition, how labels in images and tables shall be designed.

Library of Congress Control Number **LCCN** (formerly Library of Congress Catalog Card Number) for publications which are registered by the American National Library

A **legend** is a bulky explanation for tables and figures, which is delivered to the reader in addition to the figure subtitle. A legend is always located *below* the table or figure. Sometimes it appears in a box. In English the term legend is also used as a synonym for figure subtitle and used in opposition to a figure caption (above the figure), which we call figure title in this book to distinguish it from the legend.

lpi (**lines per inch**) . 6, 4 and 3 lpi means line spacing 1, 1½ and 2. This specification of the line spacing is from the type writer and matrix printer age, today it is not so usual any more.

A **macro** is an abbreviated command in a computer program, which substitutes entering several other commands (or characters in a word-processing program).

Majuscules (**capital letters**) is a → font attribute. To write text, only capital letters are used. Text in majuscules is much harder to read than text with capital *and* small letters.

Minuscles. Small letters, opposite to → majuscles.

Multimedia is the combination of text, tables, images, sound and video sequences (including computer animations) to a new form of information display. If a person perceives such information, several senses are addressed at the same time. This improves the amount of learning and remembering.

The **non-breaking space** (**NBSP**) is e. g. used between the components of a multi-part abbreviation or between an abbreviated title and last name. It creates a fixed (word) gap, which may be smaller than between ordinary words in case of left- and right-justified text, and it prevents an automatic word wrap or line break resp. at that position, so that the abbreviations and names like „i.e., e.g., Dr. Minor" stay together either on the old line or on the new one.

A **non-breaking hyphen** (NBH) is a hyphen, where the word-processing program will not change the lines, under no circumstances. This prevents, that in words which are combined with an abbreviation, the abbreviation stands alone on the old line. Example: „X-ray".

OCR (**Optical Character Recognition**) is a function when scanning. A software reads the scanned pages and converts the pixels to editable characters.

PDF (**Portable Document Format**) is a page description language, which has been defined by the company Adobe. The line and page break remains unchanged, i. e. the reader sees the pages on his computer in exactly the same way as the author has created them → Acrobat Reader.

A **pictogram** is a pictorial symbol → icon.If an image is saved in a **pixel graphics** format, the image information contains single image points (pixels). For each pixel it is saved, where it is and which color it has. Therefore the file size quickly rises, if the image area rises (opposite: → vector graphics).

PostScript (**PS**) is a page description language, which has been defined by the company Adobe. Nearly all current printers work with PostScript. Since Adobe take license fees for the screen display of PS files and distribute the → Acrobat Reader for PDF files free-of-charge, PDF has established as de-facto standard for the exchange of formatted documents.

Printed area is the area of a printed page, where „printing ink", i. e. texts, figures, headers and footers, tables etc. are or may be. In multi-column typesetting the printing area for each column is limited by → white space.

A **proportional font** is a font with variable character spacing (Times New Roman, Arial etc.). That means, to put it simply, that in a text written with a proportional font the distance from the end of one letter to the beginning of the next letter is constant (opposite: → fixed spacing font).

protected hyphen → non-breaking hyphen

protected space → non-breaking space

Point (pt): Unit of length in the graphical industry, e. g. used for character height and line thickness.

With a **rasterizing film** you can disseminate a → half tone image during copying or a negative during the enlargement into more or less coarse pixels.

Rasterizing is the dissemination of a halftone image into pixels, see above, but it is also the filling of surfaces or the background of text boxes and table cells with gray or color of different intensity. In word-processing programs this is sometimes called shading.

Reading aids are all lists/indexes and labels of a document, which exceed the pure text with figures and tables, i. e. all types of lists/indexes, footnotes, marginalia, register markings, headers and footers as well as column headings.

RFID (Radio **F**requency **Id**entification) means identification of things and creatures (e. g. containers, dogs) by means of electromagnetic waves radio chip transmits. RFID-Chips are also hidden in the spine of books in the library.

For **scalable fonts** you can select the font size in the word-processing program. It is specified in the typographic unit → point (pt).

Serial Item and Contribution Identifier **SICI** for articles and contributions to periodicals

Serifs are the small lines at the ends of the letters, e. g. from the Times family. While reading they help the eye to hold the line.

Soft hyphen → hyphenation proposal

Small caps is a font attribute using no small letters, but only normal-sized and a little smaller capital letters. See also → majuscles.

Space character → non-breaking space

The **structure** contains every document part number and heading, but no page numbers. It contains the logic of the contents, the "backbone". It is an intermediate result and grows with further writing of the Technical Report via the states 4-point-structure and 10-point-structure up to the final detailed structure.

The **Style Guide** is a collection of certain notations, technical terms and layout rules for a larger document (from about 20 pages on). It helps, that within a larger work the same items are always expressed (terminology) or displayed (layout) in the same way, i.e. that the work is *consistent* in itself.

A table of contents (ToC) contains for each document part the document part number, document part heading and page number and allows to quickly find chapters, sub-chapters, sections etc.

In **text images** text and structuring or connecting lines/arrows/boxes are aligned image-like, so that understanding the message is facilitated and the remembering of the contents is improved. Text images are often used on slides → DTV.

Text formatting → Layout.

A **text table** is a table, which predominantly contains text.

Typography is the positioning of printing ink on the paper. It is distinguished between macro-typography (on text or page level) and micro-typography (on character level).

Unicode is an international computer code for calligraphic and text symbols from all known languages, writing cultures and character systems on the earth (see also → ASCII). Unicode shall eliminate different incompatible codes in different countries or cultures. The Unicode character set is standardized in ISO 10646. One character needs 21 bit space.

If an image is saved in a vector graphics format, the image information contains scalable geometry information (e.g. center point coordinates and radius of a circle). For these geometry objects also line and fill color, line type, filling pattern etc. are saved. The file size of vector graphics is much smaller than the file size of a → pixel graphic showing the same items.

A **viewer** is a program to look at text and graphics files.

White space is a white area on the page, where there are no alphanumerical symbols, e. g. the spare line between two paragraphs or the white space between table cells and rows (if the cells are not limited by lines).

Zip file → compression.

Cited and Recommended References

Books, Articles etc.

Since all but one cited resources in this section of chapter 7 are written in German and anyone who wants to use them must be able to understand German, I have not adopted the list of references to ISO 690 and I did not translate the bibliographical data into English.

Ammelburg, G.: Rhetorik für den Ingenieur. 5. Aufl. Düsseldorf: VDI-Verlag, 1991

Baker, W. H.: How To Produce and Communicate Structured Text. In: Technical Communication. 41 (1994), p. 456–466

Bargel, H.-J.; Schulze, G.: Werkstoffkunde. 12. Aufl. Berlin: Springer Vieweg, 2018

Brändle, M. et al. Praxisleitfaden Betriebsanleitungen. tekom (Hrsg.), Stuttgart, 4. Aufl. 2014

Brehler, R: Modernes Redetraining. Niedernhausen/TS: Falken, 1995

Decker, K.-H.: Maschinenelemente: Tabellen und Diagramme. 20. Aufl. München: Carl Hanser, 2018; außerdem ist ein Aufgabenbuch und ein Formelbuch erhältlich

Dudenredaktion (Hrsg.): Duden – die deutsche Rechtschreibung. 27. Aufl. Berlin: Dudenverlag, 2017

Erdmann, E. et al. Praxisleitfäden: Regelbasiertes Schreiben - Englisch für deutschsprachige Autoren. tekom (Hrsg.), Stuttgart, 2. Auflage 2017

Fritz, A. H.; Schulze, G.: Fertigungstechnik. 12. Aufl. Berlin: Springer Vieweg, 2018

Fritz, A. (Hrsg.); Hoischen, H. (Begr. d. Werks): Technisches Zeichnen: Grundlagen, Normen, Beispiele, darstellende Geometrie. 36. Aufl. Berlin: Scriptor, 2018

Gabriel, C.-H. et al. Richtlinie zur Erstellung von Sicherheitshinweisen in Betriebsanleitungen. tekom (Hrsg.), Stuttgart, 2005

Grote, K.-H., Feldhusen, J. (Hrsg.): Dubbel – Taschenbuch für den Maschinenbau. 24. Aufl. Berlin, Heidelberg: Springer Vieweg, 2014

Grünig, C.; Mielke, G.: Präsentieren und überzeugen. Planegg/München: Haufe, 2004

Hartmann, M., Ulbrich, B., Jacobs-Strack, D.: Gekonnt vortragen und präsentieren. Weinheim: Beltz Verlag, 2004

Hering, H.: Verbesserung des Arbeitsschutzes beim Schweißen durch Einsatz brennerintegrierter Absaugdüsen: Effektivität und Qualitätssicherung. Große Studienarbeit, betreut vom Institut für Fabrikanlagen der Universität Hannover und dem Heinz-Piest-Institut für Handwerkstechnik an der Universität Hannover, 1987

© Springer-Verlag GmbH Germany, part of Springer Nature 2019

H. Hering, *How to Write Technical Reports*,

https://doi.org/10.1007/978-3-662-58107-0

Hering, H.: Berufsanforderungen und Berufsausbildung Technischer Redakteure: Verständlich schreiben im Spannungsfeld von Technik und Kommunikation. Dissertation, Universität Klagenfurt, 1993

Hering, L.: Computergestützte Werkstoffwahl in der Konstruktionsausbildung: CAMS in Design Education. Dissertation, Universität Klagenfurt, 1990

Hering, L; Hering, H.; Kurmeyer, U.: EDV für Einsteiger. 2. Aufl. Hemmingen, 1995

Hering,L.; Hering, H.; Köhler, N.: Der TEXTdesigner: Computergestützte Analyse und Optimierung der Verständlichkeit von Sachtexten aller Art. Computerprogramm und Handbuch. Hemmingen, 1994

Herrmann, P.: Reden wie ein Profi. München: Orbis, 1991

Hermann, U.; Götze, L.: Die deutsche Rechtschreibung. Gütersloh, München: Bertelsmann Lexikon Verlag 2002

Holzbaur, U; Holzbaur, M.: Die wissenschaftliche Arbeit. München: Hanser, 1998

Horn, J.: Urheberrecht beim Einsatz neuer Medien in der Hochschullehre. Oldenburg: OLWIR Verlag, 2007

Ilzhöfer, V.: Patent-, Marken- und Urheberrecht – Leitfaden für Ausbildung und Praxis. 10. Aufl. München: Vahlen, 2018

Jung, H.: Versammlung und Diskussion. München, Goldmann, 1980

Klein, M.: Einführung in die DIN-Normen. hrsg. vom DIN, bearb. v. K. G. Krieg 14. Aufl. Wiesbaden: B. G. Teubner und Berlin: Beuth, 2008

Kurz, U., Wittel, H; Böttcher, P. (Begr. d. Werks): Böttcher/Forberg, Technisches Zeichnen – Grundlagen, Normung, Übungen und Projektaufgaben. 26. Aufl. Wiesbaden: Springer Vieweg, 2014

Labisch, S.; Weber, Chr.: Technisches Zeichnen: Selbstständig lernen und effektiv üben. 4. Aufl. Wiesbaden: Springer Vieweg, 2013

Marks, H.E.: Der technische Bericht: Ein Leitfaden zum Abfassen von Fachaufsätzen sowie zum Vorbereiten von Vorträgen. 2. Aufl. Düsseldorf: VDI-Verlag, 1975

Melezinek, A.: Unterrichtstechnologie. Wien, New York: Springer, 1982

Melezinek, A.: Ingenieurpädagogik – Praxis der Vermittlung technischen Wissens. 4. Aufl. Wien: Springer, 1999

N. N.: Intensivkurs Neue Rechtschreibung. Köln: Serges Medien, 1998

Nordemann, W.; Vinck. K.; Hertin, P.W.: Urheberrecht: Kommentar zum Urheberrechtsgesetz, Verlagsgesetz, Urheberrechtswahrnehmungsgesetz. 10. Aufl. Stuttgart: Kohlhammer, 2008

Rehbinder, M; Hubmann, H.: Urheberrecht. 17. Aufl. München: Beck, 2015 (aktualisierte bibliografische Angaben)

Reichert, G. W.: Kompendium für Technische Anleitungen. 6. Aufl. Leinfelden-Echterdingen: Konradin, 1989

Reichert, G. W.: Kompendium für Technische Dokumentationen. 2. Aufl. Leinfelden-Echterdingen: Konradin, 1993

Seifert, J. W.: Visualisieren – Präsentieren – Moderieren. 30. Aufl. Offenbach: Gabal, 2011

Theisen, M.: Wissenschaftliches Arbeiten. 17. Aufl. München: Vahlen, Franz, 2017

Thiele, A.: Überzeugend Präsentieren. 2. Aufl. Berlin: Springer, 2000

Wittel, H. et al.: Roloff/Matek Maschinenelemente. 23 Aufl. Wiesbaden: Springer Vieweg, 2017

Standards, Guidelines etc.

German standards are listed with German references, international standards are listed with English references. If the references contain an entry „mehrere Teile (oder Blätter)" or „several parts (or sheets)", the standard consists of at least two parts (or sheets), so that it is not possible to list the year of publication, because the parts (or sheets) of the standard or guideline have been published in different years.

DIN, Deutsches Institut für Normung (Hrsg.): Berlin: Beuth
DIN 108 Diaprojektoren und Diapositive, mehrere Teile
DIN 406-10:1992-12 Technische Zeichnungen; Maßeintragung; Begriffe, allgemeine Grundlagen
DIN 406-11:1992-12 Technisches Zeichnen – Maßeintragung, Teil 11: Grundlagen der Anwendung
DIN 461:1973-03 Graphische Darstellungen in Koordinatensystemen
DIN 616:2000-06 Wälzlager, Maßpläne <Anm. d. Verf.: Anschlussmaße von Wälzlagern>
DIN 623-2:2000-06 Wälzlager; Grundlagen; Teil2: Zeichnerische Darstellung von Wälzlagern
DIN 824:1981-03 Technische Zeichnungen – Faltung auf Ablageformat
DIN 1301:2010-10 Einheiten, mehrere Teile, u. a. Teil 1: Einheitennamen, Einheitenzeichen
DIN 1302:1999-12 Allgemeine mathematische Zeichen und Begriffe
DIN 1303:1987-03 Vektoren, Matrizen, Tensoren – Zeichen und Begriffe
DIN 1304-1:1994-03 Formelzeichen; Allgemeine Formelzeichen
DIN 1313:1998-12 Größen <Anm. d. Verf.: enthält auch Einheitensysteme und Gleichungen>
DIN 1338:2011-03 Formelschreibweise und Formelsatz
DIN 1421:1983-01 Gliederung und Benummerung in Texten
DIN 1422-1:1983-02 Veröffentlichungen aus Wissenschaft, Technik, Wirtschaft und Verwaltung; Teil 1: Gestaltung von Manuskripten und Typoskripten
DIN 1426:1988-10 Inhaltsangaben von Dokumenten; Kurzreferate, Literaturberichte
DIN 1460:1982-04 Umschrift kyrillischer Alphabete slawischer Sprachen
DIN 2340:2009-04 Kurzformen für Benennungen und Namen
DIN 5007:2005-08 Ordnen von Schriftzeichenfolgen (ABC-Regeln)
DIN 5008:2011-04 Schreib- und Gestaltungsregeln für die Textverarbeitung
DIN 5473:1992-07 Logik und Mengenlehre – Zeichen und Begriffe
DIN 5478:1973-10 Maßstäbe in graphischen Darstellungen
DIN 5483:1982-03 Zeitabhängige Größen; Formelzeichen
DIN 16511:1966-01 Korrekturzeichen
DIN 19045-3:1998-12 Projektion für Steh- und Laufbild – Teil 3: Mindestmaße für kleinste Bildelemente, Linienbreiten, Schrift- und Bildzeichengrößen in Originalvorlagen für die Projektion
DIN 31051:2012-09 Grundlagen der Instandhaltung (auch DIN 31051:2018-09, Entwurf)
DIN 31623-1:1988-09 Indexierung zur inhaltlichen Erschließung von Dokumenten; Begriffe, Grundlagen
DIN 31634:2011-10 Information und Dokumentation – Umschrift des griechischen Alphabets
DIN 31635:2011-07 Information und Dokumentation – Umschrift des arabischen Alphabets für die Sprachen Arabisch, Osmanisch-Türkisch, Persisch, Kurdisch, Urdu und Paschtu
DIN 31636:2011-01 Information und Dokumentation – Umschrift des hebräischen Alphabets (auch DIN 31636:2018-04 - Entwurf)
DIN 31638:1994-08 Bibliografische Ordnungsregeln
DIN 32520 Grafische Symbole für die Schweißtechnik, mehrere Teile
DIN 55301:1978-09 Gestaltung statistischer Tabellen
DIN 66001:1983-12 Informationsverarbeitung, Sinnbilder und ihre Anwendung
DIN 66261:1985-11 Sinnbilder für Struktogramme

DIN EN 82079:2018-05; VDE 0039-1:2018-05 (Entwurf) Erstellen von Anleitungen; Gliederung, Inhalt und Darstellung – Teil 1: Allgemeine Grundsätze und ausführliche Anforderungen; Text Deutsch und Englisch

DIN ISO 128 Technische Zeichnungen; Allgemeine Grundlagen der Darstellung, mehrere Teile, u. a. zu Linien, Ansichten, Flächen in Schnitten und Schnittansichten

DIN ISO 1101:2017-09 Geometrische Produktspezifikation (GPS) - Geometrische Tolerierung - Tolerierung von Form, Richtung, Ort und Lauf

DIN ISO 2768 Allgemeintoleranzen ... ohne einzelne Toleranzeintragung, mehrere Teile aus 1991

DIN ISO 5456 Technische Zeichnungen; Projektionsmethoden, mehrere Teile aus 1998

VDI, Verein Deutscher Ingenieure (Hrsg.): Düsseldorf, Berlin: Beuth

VDI 2222-2225 Konstruktionsmethodik, mehrere Blätter

VDI 2244:1988-05 Konstruieren sicherheitsgerechter Erzeugnisse

VDI 4500 Blatt 1:2006-06 Technische Dokumentation – Benutzerinformation, Begriffsdefinitionen und rechtliche Grundlagen

VDI 4500 Blatt 2:2006-11 Technische Dokumentation – Organisieren und Verwalten

VDI 4500 Blatt 3:2006-06 Technische Dokumentation – Erstellen und Verteilen elektronischer Ersatzteilinformationen

VDI 4500 Blatt 4:2011-12 Technische Dokumentation – Dokumentationsprozess – Planen, Gestalten, Erstellen

VDI 4500 Blatt 6:2017-11 (Entwurf) Technische Dokumentation – Dokumentationsprozess – Publizieren

ISO, International Organisation for Standardization (Hrsg.)

ISO 4:1997-12 Information and documentation - Rules for the abbreviation of title words and titles of publications

ISO 8:1977-09 Documentation; Presentation of periodicals

ISO 9:1995-02 Information and documentation - Transliteration of Cyrillic characters into Latin characters - Slavic and non-Slavic languages

ISO 128-1:2003-02 Technical drawings - General principles of presentation - Part 1: Introduction and index

ISO 128-20:1996-11 Technical drawings - General principles of presentation - Part 20: Basic conventions for lines

ISO 128-21:1997-03 Technical drawings - General principles of presentation - Part 21: Preparation of lines by CAD systems

ISO 128-22:1999-05 Technical drawings - General principles of presentation - Part 22: Basic conventions and applications for leader lines and reference lines

DIN ISO 128-24:1999-12 Technical drawings - General principles of presentation - Part 24: Lines on mechanical engineering drawings (ISO 128-24:1999)

ISO 128-30:2001-04 Titel (Deutsch): Technical drawings - General principles of presentation - Part 30: Basic conventions for views

ISO 128-44:2001-04 Technical drawings - General principles of presentation - Part 44: Sections on mechanical engineering drawings

ISO 233:1984-12 Information and documentation; transliteration of Arabic characters into Latin characters; part 2: Arabic language; simplified transliteration

ISO 259-2:1994-12 Information and documentation - Transliteration of Hebrew characters into Latin characters - Part 2: Simplified transliteration

DIN ISO 690:2013-10 Information and documentation - Guidelines for bibliographic references and citations to information resources (ISO 690:2010)

ISO 832:1994-12 Information and documentation - Bibliographic description and references - Rules for the abbreviation of bibliographic terms

ISO 843:1997-01 Information and documentation - Conversion of Greek characters into Latin characters

ISO 1101:2017-02 Geometrical product specifications (GPS) - Geometrical tolerancing - Tolerances of form, orientation, location and run-out

ISO 2145:1978-12 Documentation – Numbering of divisions and subdivisions in written documents

ISO 2710-1:2017-11 Reciprocating internal combustion engines - Vocabulary - Part 1: Terms for engine design and operation

DIN EN ISO 3098-1:2015-06 Technical product documentation - Lettering - Part 1: General requirements (ISO 3098-1:2015)

DIN EN ISO 3098-2:2000-11 Technical product documentation - Lettering - Part 2: Latin alphabet, numerals and marks (ISO 3098-2:2000)

DIN EN ISO 3098-3:2000-11 Technical product documentation - Lettering - Part 3: Greek alphabet (ISO 3098-3:2000)

DIN EN ISO 3098-4:2000-11 Technical product documentation - Lettering - Part 4: Diacritical and particular marks for the Latin alphabet

DIN EN ISO 3166-1:2014-10 Codes for the representation of names of countries and their subdivisions - Part 1: Country codes (ISO 3166-1:2013)

DIN EN ISO 4762:2004-06 Hexagon socket head cap screws (ISO 4762:2004)

ISO 5456-1:1996-06 Technical drawings - Projection methods - Part 1: Synopsis

ISO 5456-2:1996-06 Technical drawings - Projection methods - Part 2: Orthographic representations

ISO 5456-3:1996-06 Technical drawings - Projection methods - Part 3: Axonometric representations

ISO 5456-4:1996-06 Technical drawings - Projection methods - Part 4: Central projection

ISO 5776:2016-04 Graphic technology - Symbols for text proof correction

ISO 6410-3:1993-12 Technical drawings; screw threads and threaded parts; simplified representation

ISO 7098:2015-12 Information and documentation - Romanization of Chinese

ISO 7144:1986-12 Documentation; Presentation of theses and similar documents

Internet-Links, Newsletter, and Forums

http://dnb.d-nb.de catalog of the German national library

www.mvb-online.de MVB Marketing- und Verlagsservice des Buchhandels GmbH creates the VLB

www.vlb.de Verzeichnis lieferbarer Bücher VLB (list of deliverable books), register your publications there.

www.german-isbn.org, if you want to apply for an ISBN

www.vgwort.de In order to get a financial portion, if libraries buy your publication or if readers copy your publication, you must register your publication with the VG Wort.

www.beuth.de and www.iso.org to purchase and look up standards

www.din5008.de or resp. www.tastschreiben.de: DIN 5008 in the exact wording, and a detailed representation of the rules for the new German orthography

www.duden.de/rechtschreibpruefung-online: spell checker (max. 800 characters)

dict.leo.org and www.dict.cc: dictionaries for different translation directions

www.systranbox.com/systran/box: translation of short texts in different translation directions

http://www.google.de/language_tools?hl=de: translation of short texts in different translation directions

http://de.wikipedia.org/wiki/Liste_von_Konjunktionen_im_Deutschen

http://grammar.yourdictionary.com/parts-of-speech/conjunctions/conjunctions.html

www.thepunctuationguide.com punctuation rules for texts in American English
www.chicagomanualofstyle.org grammar, usage, citing of texts, valid for texts in American English
http://openpdf.com/ebook/iso-690-1-pdf.html
HTML reference SelfHTML by Stefan Münz: https://wiki.selfhtml.org/wiki/Startseite
HTML reference by w3schools.com http://www.w3schools.com/tags/
Online HTML editing: http://htmledit.squarefree.com/
HTML editor Phase 5 (free-of-charge): http://www.phase5.info.
HTML editor HTML-kit 292 (free-of-charge): http://www.htmlkit.com/
HTML editor for responsive web design (free-of-charge): http://www.coffeecup.com/free-editor/
Editing formulas in HTML format www.mathe-online.at/formeln

Index

A
abbreviations, 131
ABC rules, 102
abstract, 50
accentuation, 136
acknowledgements, 50
adhesive binding, 157
Anhang
    Dateien, 55
annexes, 50
appendix
    catalogues, 53
    manufacturer catalogues, 53
    one chapter, 51
    page numbering, 51
    separately bound, 155
    several chapters, 51
    structure, 51
    structure acc. to ISO 7144, 51
    title leaf, 54
assembly dimensions, 86
assembly drawing, 86
author names, 109

B
backbone, 11
back-up copies of files, 165, 167
bibliographical data, 99, 107
    abbreviations, 111
    acc.to ISO 690 and ISO 690-2, 111
    author names, 109
    elements, 111
    structure, 111
bibliography, 107
bill of materials, 26, 85
binding
    what to do before, 152
binding type
    booklets of any type, 154
    brochures, 154
    cold adhesive binding, 157

    comb binding, 156
    file binder, 155
    filing fastener, 155
    folder, 154
    handout, 154
    hardcover binding, 157
    hot adhesive binding, 157
    journals, 154
    overview, 152
    paper-clip, 154
    plastic folder, 154
    presentation documentation, 154
    protocol, 154
    ring binder, 155
    saddle-stitching, 154
    script, 154
    spring binder, 155
    spring strip, 155
    staple, 154
    staple binding, 157
    wire-O binding, 156
block format, 108
book
    bibliographical data, 112
book pile, 164
bullet lists, 135

C
cabinet projection, 84
CAD programs, 88
capital letters, 136
cavalier projection, 84
central projection, 84
chapter number 0, 14
charisma, 177, 179, 211, 217
chart, 77
citation, *see* literature citation
    figures, 105
    position in the sentence, 105
    selection of brackets, 100
    tables, 105

© Springer-Verlag GmbH Germany, part of Springer Nature 2019
H. Hering, *How to Write Technical Reports*,
https://doi.org/10.1007/978-3-662-58107-0

citing
  book, 112
  book with CD-ROM or DVD, 112
  book with video cassette, 112
  brochure, 113
  CD, 98
  CD-ROM, 98
  computer program, 98
  host document, 112
  information from data networks, 98, 115
  intellectual property rights, 113
  journal, 112
  manuafacturer documents, 113
  monograph, 112
  patent, 113
  publication by an institution, 113
  publications by companies, 113
  radio transmission, 98
  record, 98
  source from the internet, 113
  special cases, 114
  trade mark, 113
  TV film, 98
  video film, 98
clarity of the text, 120
CMYK, 92
cold adhesive binding, 157
colors
  possible applications, 193
comb binding, 156
computations, 123
connecting sentence, 120
consistency
  layout, 30
  terminology, 30
consistency of terms
  in literature citations, 95
consistent names, 131
contact preparations, 213
contacting the audience, 213
contents, 41
coordinate axes
  scale, 78
coordinate system, 77
copier
  recycling paper, 147
copy originals, 146
  additional labels, 148
  glueing in figures/tables, 147
  only the original with colors, 148
copy shop
  working together with the ∼, 133
copyright laws
  information from data networks, 115

corporate design, 30, 33, 132
corporate identity, 36
correction symbols, 142
cover sheet, 35
  faults, 36
  inner cover, 35
  outer cover, 35
  required information, 35
curriculum vitae (CV), 56
curve flow, 77
cutaway drawing, 85

D

data networks
  citing information, 115
declaration in lieu of an oath, 50
definition
  Technical Report, 1
design description, 26, 122
design methodology, 61
diagram, 77
  change of the optical impression, 80
  choosing the scale density, 80
  complete labeling, 77
  distinguishing between several curves, 78
  error tolerance zone, 80
  interrupted axes, 78
  measured point symbols, 78
  qualitative relationship, 80
  quantitative relationship, 78
  ruled lines, 78
  scaling of ruled lines, 78
  tolerances, 80
  types and fields of application, 81
diagram axes
  scale, 78
didactic reduction, 74, 95
didactic-typographic visualisation, 88
didactic-typographic visualization, 69
digital image
  distortion, 92
  format, 92
  resolution, 92
digital photo, 88
  emphasis and accentuation, 95
  rules for the design, 93
  simplifications, 95
  undesired reflection of flashlight, 93
digital printing, 133
dimetric projection, 84
document layout, 132
document part heading, 13, 14
  design and layout rules, 16

formatting style, 17
  layout, 134
  logic, 14
document part number, 14
  design and layout rules, 16
  logic, 14
drawing, 86
  folding, 155
  set of ~, 26
DTV, 69, 88

E
editor, 109
  working together with the ~, 133
equation number, 123
equations, 123
error tolerance zone
  display in diagrams, 80
evaluation tables, 65
  legend, 65
  opposite point values, 67
  parallel point values, 67
  weighting factors, 66
experimental work, 27
exploded view, 85

F
facial expression, 179, 217
figure numbering, 75
figure subheading
  formatting style, 76
  layout, 134
  more than one line, 76
  placement, 75
  precise location references, 76
  re-numbering, 77
  several small figures, 76
figures, 70
  annex material in the appendix, 70
  apperception, 129
  basic design rules, 72
  colors hard to copy, 148
  continuation, 75
  create associations, 71
  cross-references in the text, 71, 76
  drawn with stencils, 147
  emphasis and accentuation, 95
  explain abstract information, 71
  figure subheadings/titles, 75
  freehand drawing, 147
  from the internet, 88
  glueing straight with ruler, 147

graph paper, 85
  isometric drawing paper, 85
  leave enough space for gluing in the figures,
    95
  manual creation, 81, 85
  manually drawn, 147
  note of reference, 77
  numbering, 75
  placement, 77
  recognition, 82
  scanned, 88
  shape laws
  simplifications, 95
  simplify reality, 71
  swing-out creation, 148
  templates, 85
  transparency paper, 147
figures (list of ~), 52
file binder, 155
file organization, 165
filing fastener, 155
final printout, 146
  leading dots in table of contents, 45
  paper quality, 147
folder, 154
font size, 136
font type, 136
footers
  layout, 134
footnotes
  for comments, 128
  for literature citations, 128
  in tables, 129
foreword, 14, 50
formatting style, 139
  document part heading, 17
  figure subheading, 76
  table heading, 60
formula editor, 124
formulas, 123
  as normal text, 124
  computations of loads, 127
  definition, 123
  in HTML documents, 124
  legend, 126
  parts, 123
  relationship to the text, 125
  with LATEX, 125
  with Word, 124
front cover, 35
  faults, 36
  minimum information, 40
  placement of information, 41
further reading, 107

G

gestures, 179, 206, 217
GIF files, 92
glossary, 55
graphics
    systematic structure, 71
graphics programs
    advantages, 89
    comparison with manual creation, 90
    disadvantages, 89

H

halftone image, 90
hard disk organisation, 165
hardcover binding, 157
headers
    layout, 134
host document
    bibliographical data, 112
hot adhesive binding, 157
HTML creation, 149
hyphen
    protected hyphen, 139
    soft hyphen, 139
    structuring words, 139
hyphenation
    influencing ∼, 139
    proposals, 139
    with special characters, 139
hyphens in composite nouns, 131
hyphens in technical terms, 131

I

images
    emphasis and accentuation, 95
    simplifications, 95
indentations
    of text, 135
index, 56
information (definition), 177
information from data networks
    bibliographical data, 115
inner cover, 35
instructions for operation, 28
instructions for use, 28
intellectual property rights
    bibliographical data, 113
internet
    bibliographical data, 113
internet Explorer
    export favorites, 169
introduction, 48

introductory sentence, 120
isometric projection, 84

J

joining chamfers, 86
jotter (project notebook), 29
journal
    bibliographical data, 112, 114
JPG files, 92

L

LATEX, 125
layout, 132
    consistency, 30
    document part heading, 134
    figure subheading, 134
    footers, 134
    headers, 134
    line break, 138
    page make-up, 138
    page margins, 132
    page numbers, 133
    table heading, 134
leading dots
    in lists and indexes, 43
    in table of contents, 43
Lecture, see presentation
lecture (definition), 176, 178
left- and right justified, 135
left justified, 135
legend, 65
library work, 162
    check copies, 163
    coin box, 162
    copy card, 162
    order of work steps, 163
    paper clips, 162, 163
    photocopying, 163
    preparing the list of references, 163
line break
    layout, 138
line drawing, 90
line spacing, 135
Linux
    Kanopix-CD, 125
    Knoppix-CD, 125
list of abbreviations, 53
list of figures, 52
list of references, 49, 107
    abbreviations, 111
    acc. to ISO 690 and ISO 690-2, 111
    author names, 109

block format, 108
capitular list of references, 108
classic form, three columns, 108
example, classic, 3 columns, 117
sequence of references, 102
sorting by alphabetical order, 102
space-saving, two columns, 108, 118
structured, unstructured, 107
list of requirements, 61
list of tables, 52
list of used formulas and units, 53
lists
    nested, 135
literature, 107
literature citation, 96
    analogous citation, 104
    comment, 104
    definition of terms, 96
    figures, 105
    image from the internet, 107
    literal literature citation, 103
    parts, 99
    reasons, 97
    rules, 99
    source from the internet, 113
    tables, 105
    tasks, 96
    types, 98
literature number, 101
literature sources
    identification, 101
logic, 14
    document part heading, 14
    document part number, 14
    structuring principles, 24
logic of language, 14
logic of the sequence of thoughts and worksteps, 14

M
macrotypography, 00
main target of the report, 18
main topic of the report, 18
manual, 28
manufacturer catalogues, 53
manufacturer documents
    bibliographical data, 113
material collection, 57
    book pile, 164
    notes, 165
measured point symbols, 78
microtypography, 234
mimic, 217

monograph
    bibliographical data, 112
morphological box, 61
    combination of several solutions of sub functions, 64
    numbering the solutions of the sub functions, 64
    numbering the sub functions, 64
    subdivision of a sub function, 64
    verbal evaluation, 62
mounted part drawing, 26

N
names
    according to the function, 121
    unique and consistent, 131
nested lists, 135
nonverbal communication, 179
notebook, 29

O
occasional speech, 178
oral communication
    advantages and disadvantages, 179
ordering the material, 57
orthogonal projection, 84
outer cover, 35

P
page break, 138
page layout, 132
page make-up
    layout, 138
page margins
    layout, 132
    space for binding, 133
page numbering
    according to ISO 2145, 42
    according to ISO 7144, 43
page numbers
    layout, 133
paper organization, 164
paper-clip, 154
paperwork
    book pile, 164
    notes, 165
    report binder, 164
    to-do-list, 165
paragraph layout, 134
paragraphs
    spacing, 134

part names, 121
　unique and consistent, 131
passive voice, 121
patent
　bibliographical data, 113
PDF creation, 149
personal working methodology, 168
　buy materials in time, 168
　export internet favorites, 169
　go-on-here marking, 170
　not-yet-ready marking, 169
　style guide, 169
　teamwork, 169
　time estimation, 168
　writing the text, 170
perspective drawing, 84
　advantages, 85
　cutaway drawing, 85
　disadvantages, 85
　exploded view, 85
perspective projections
　comparison, 84
persuasive presentation, 178
photo, 88
　emphasis and accentuation, 95
　exposure with rasterizing film, 94
　rules for the design, 93
　simplifications, 95
photocopy, 88
　emphasis and accentuation, 95
　simplifications, 95
　undesired terms, 94
　with photo key, 96
　with rasterizing film, 96
plastic folder, 154
PNG files, 92
preface, 14, 50
present (definition), 176
presentation
　appropriate pointing, 215
　audience, target group, 183
　backbone, 187
　colour code for manuscript cards, 208
　contact preparations, 213
　contacting the audience, 213
　corporate identity, 195
　creation, 191
　design, 188, 191
　difference to Technical Report, 176
　documentation, 209
　dramaturgy, 187, 191
　example—backbone, 187
　example—presentation framework, 185
　example—presentation target, 185

example—structure, 187
example—visualization, 201
figure title, 199
framework slides, 193
handout, 209
intermediate question, 215
laser pointer, 215
last changes, 211
manuscript, 206
manuscript cards, 206
material collection, 186
mazy, 189
necessary worksteps, 180
new, 189
own title, 213
personal preparation, 211
pointer, 215
preparation of the meeting room, 212
preparations in the room, 211
presentation framework, 181
presentation targets, 181, 184
presentation title, 182
presentation type, 183
prominent people, 213
real intermediate questions, 215
review, 216
slide heading, 199
structure, 187
structure design plot, 196
summarizing, 196
technical preparation, 211
telescopic pointer, 215
third-rule, 188
title of prominents, 213
trial presentation, 210
trisection in a technical presentation, 189
trisection in biology and technology, 189
trisection in example presentation, 192
unreal intermediate questions, 216
updating the presentation, 211
visualization, 192, 197
well-known, 189
working out the details, 196
presentation (definition), 176
presentation targets, 176, 177
presentation types, 177, 178
　lecture, 178
　occasional speech, 178
　persuasive presentation, 178
　technical presentation, 178
printer
　working together with the ~, 133
printer driver, 133
printing area, 129, 132

project notebook, 29
projection
    cabinet projection, 84
    cavalier projection, 84
    central projection, 84
    diametric projection, 84
    isometric projection, 84
    three-plane-projection, 84
prolog, 14
proof-reading, 142
protected hyphen, 139
publication, 98
publication by an institution
    bibliographical data, 113

R
rasterizing film
    for copier, 96
    for enlarging negatives, 94
recycling paper, 147
reference, 99
report binder, 164
    structure, 164
report checklist, 140
rhetoric, 222
ring binder, 155
rough design description, one concept variant,
    26
rough design description, several concept vari-
    ants, 25
ruled lines, 78
    scaling, 78

S
saddle-stitching, 154
safety notes, 28
scanned image
    emphasis and accentuation, 95
    simplifications, 95
scheme, 77
    standardized symbols, 77
secondary citation, 102
section drawing, 86
section lining, 87
sentence level
    understandability of text, 129
sentence structure, 130
serifes, 136
set of drawings, 26
shape laws, 00
simplified technical drawing, 83
single part drawing, 86

sketch
    figure subheading, 83
    illustration of computation, 82
    labeling, 83
    simplified technical drawing, 82
slides
    basic layout, 192
    footers, 195
    including text and images, 194
    readability check, 193
    slide master, 192
    structure on the left, 195
    transparency assurance, 193
small caps, 136
soft hyphen, 139
source from the internet
    bibliographical data, 113
space
    non-breaking space, 139
speech forms, 176
spelling
    consistent usage, 30
spiral binding
    with plastic spiral/comb, 156
    with wire spirals, 156
spring binder, 155
spring strip, 155
standardized terms, 122
staple, 154
staple binding, 157
stapling
    what to do before, 152
statement (definition), 177
structure, 11
    10-point-structure, 18
    4-point-structure, 18
    backbone, 11
    check for completeness, 18
    check for correctness, 18
    design report, 19
    detailed structure, 18
    distinction from table of contents, 11
    report about a computer network enhance-
        ment, 21
    report about executed measurements, 20
    report about the development of software, 22
    rough design description, one concept vari-
        ant, 26
    rough design description, several concept
        variants, 25
    structure patterns, 25
    structuring principles, 24
    structuring principles of instructions for use,
        product-oriented or task-oriented, 28

worksteps to create it, 24
structure pattern, 25
    experimental work, 27
    instructions for operation, 28
    instructions for use, 28
    manual, 28
    rough design description, one concept variant, 26
    rough design description, several concept variants, 25
style guide, 30
subchapter number n.0, 14
subheadings, 46
summary, 48
swing-out
    creation, 148
    glueing and binding, 152

T
table heading
    formatting style, 60
    layout, 134
    more than one line, 60
    positioning, 60
table numbering, 59
table of contents, 41
    capitular table of contents, 46
    end check, 47
    final printout, 45
    indentations, 43
    leading dots, 43
    manually typed, 47
    overall table of contents, 46
    typographic design, 43
tables, 58
    annex material in the appendix, 58
    continuation, 59
    cross-references in the text, 60
    footnotes, 129
    glueing straight with ruler, 147
    note of reference, 61
    numbering, 58
    swing-out creation, 148
    table headings/titles, 58
tables (list of ~), 52
target group, 137
task, 49
teamwork, 161
technical drawing, 85
    center lines, 85
    frequent mistakes, 85
technical presentation, 178

Technical Report
    binding, 151
    copying, 151
    creating HTML and PDF, 149
    creation, 33
    digital printing, 133, 140
    parts according to ISO 7144, 34
    publishing in data networks, 149, 151
    stapling, 151
    types, 1
    work steps in the network plan, 6
technical terms, 131
tense, 121
terminology
    consistency, 30
    consistent spelling and usage, 30
    problems in cited figures, 94
terms
    standardized terms, 122
text
    added graphical elements, 69, 88
    didactic-typographic visualisation, 88
    didactic-typographic visualization, 69
    sentence structure, 130
    understandability on sentence level, 129
    understandability on text level, 128
    understandability on word level, 131
    writing style, 119
text accentuation, 136
text charts, 69, 88
text justification, 135
text level, 128
text-figure-relationship, 70
three-plane projection, 84
TIF files, 92
time estimation, 168
title, 8
    data base searches, 8
    keywords, 10
    main keywords, 8
    speech melody, 8
    subtitle, 10
    title creation, 10
title leaf, 35
    faults, 36
    font type, 37
    layout examples, 37
    minimum information, 40
    placement of information, 41
ToC (table of contents), 41
trade mark
    bibliographical data, 113
typography, 129

U

unambiguousness of text, 120
understandability
  abbreviations, 131
  figures, 72
  reader properties, 127
  sentence level, 129
  sentence structure, 130
  structuring words, 139
  technical terms, 131
  text, 119
  text level, 128
  text properties, 127
  word level, 131
undesired terms, 94
unique names, 131
Unterkapitelnummer n.0, 14

V

visualising text, 88
visualization
  animation, 200
  perceptability, 199
  readability, 199

transparency assurance, 193
visualizing text, 69
voice, 179

W

warnings, 28
wire-O binding, 156
word level, 131
word processing system, 132
working in the library, 162
working together in a team, 161
working together with the supervisor or
      customer
  checklist, 160
  presentation, 160
  structure, 160
  taking notes, 161
writing style
  consistent usage, 30
  for Technical Reports, 120
  general, 119
  understandability, 127
written communication
  advantages and disadvantages, 179

Printed in the United States
By Bookmasters